REAL SOLIDS
AND RADIATION

A. E. Hughes and D. Pooley
Atomic Energy Research Establishment
Harwell

WYKEHAM PUBLICATIONS (LONDON) LTD
(A MEMBER OF THE TAYLOR & FRANCIS GROUP)
LONDON AND WINCHESTER
1975

First published 1975 by Wykeham Publications (London) Ltd.

Cover illustration—Information storage in crystals using radiation. The photograph shows how one million bits of information can be stored on the surface of a crystal only 1·5 mm square. The information stored is actually a repeated visual pattern and a section near the centre has been deliberately erased.

ISBN 0 85109 011 7

Printed in Great Britain by Taylor & Francis Ltd.
10–14 Macklin Street, London, WC2B 5NF

1004090 9|1

Distribution:

UNITED KINGDOM, EUROPE AND AFRICA
Chapman & Hall Ltd. (a member of Associated Book Publishers Ltd.) 11 North Way, Andover, Hampshire.

WESTERN HEMISPHERE
Springer-Verlag New York Inc., 175 Fifth Avenue, New York, New York 10010.

AUSTRALIA, NEW ZEALAND AND FAR EAST (excluding Japan)
Australia and New Zealand Book Co. Pty. Ltd.,
P.O. Box 459, Brookvale, N.S.W. 2100.

INDIA, BANGLADESH, SRI LANKA AND BURMA
Arnold-Heineman Publishers (India) Pvt. Ltd.,
AB–9, First Floor, Safjardang Enclave, New Delhi 11016.

ALL OTHER TERRITORIES
Taylor & Francis Ltd., 10–14 Macklin Street, London, WC2B 5NF.

NEARLY everything we do in our lives depends in one way or another on the properties of solids, so it is not surprising that a large part of the research and development activity in modern science is devoted to the study of the solid state. The physics of solids has one of its most obvious practical products in electronic devices, the chemistry of solids in plastics, dyestuffs and fibres, and the science of metallurgy is apparent through the wide range of structural materials which are now available to the engineer. The motivation for this book arose originally through our own involvement in radiation effects in solids, which is a topic invoking a wide range of modern scientific concepts without being an accepted mainstream subject which features much in formal science courses at school or university. We therefore aimed to write an account of this field which would be intelligible to the non-specialist. However, after starting a text based on this idea, and discussing it with Brian Woolnough, it seemed to us all that some benefit would derive by broadening the subject matter. Radiation effects in a solid revolve almost exclusively around the production of defects in its atomic structure, which manifest themselves through changes in various physical properties of the material. But there are many other features in the behaviour of solids which are controlled by structural defects which are present for reasons which have nothing to do with radiation at all. In many cases the presence of defects explains why the properties of many solids with which we are very familiar, such as pottery and steels, differ from the ideal solids which are discussed in most text-books. We therefore have tried to write a book for the Wykeham series which describes how the properties of 'real' solids may be understood in terms of the atomic and electronic structure of ideal solids modified by the presence of a wide range of types of structural defect. We have kept the interaction of solids with radiation as a recurring theme, partly because this is one of the more important ways in which defects are formed and has stimulated a great deal of study of defect properties, and partly because the interaction of solids with one simple form of radiation, light, is the means by which we gain much of our practical experience of the nature of real solids. The sort of general questions we aim to answer in this book are exemplified by such as "Why are some solids electrical conductors and some insulators?", "Why are some solids transparent and others opaque?", and "Why are some solids brittle and others ductile?"

The first two chapters of the book form an introduction to the structure and properties of perfect solids, developing the ideas about electrons and atoms which are required to understand the nature of the solid state. Chapters 3 and 4 introduce the types of defect which may be present in a real solid and their influence on some of the physical properties of materials. Chapters 5 and 6 describe the interaction of solids with radiation, leading to a discussion of the production of defects by irradiation in Chapter 7. Finally Chapters 8 and 9 are devoted to some of the practical consequences of radiation effects, both where they have unwanted effects and where they are used to perform some potentially useful tasks.

Some of the material covered in this book overlaps the content of other books in the Wykeham series, and in some cases we have deliberately referred to other books rather than spell out an argument in detail. The reader should find the following texts to be especially useful companions to our volume: *Elementary Quantum Mechanics* by N. F. Mott, *Elementary Science of Metals* by J. W. Martin, *Crystals and X-rays* by H. S. Lipson, *Biological Effects of Radiation* by J. E. Coggle, *Solid State Electronic Devices* by D. V. Morgan and M. J. Hawes and *Strong Materials* by J. W. Martin.

It is a pleasure to acknowledge the help we have received from our schoolmaster-collaborator, Brian Woolnough, and from the series editor Mr. G. R. Noakes. Both have contributed in many ways towards making the text more readable. Finally we should like to thank many of our colleagues at Harwell who have contributed to our understanding of the subject matter of this book, and our families who have put up with many hours of silence and detached contemplation during its writing.

Harwell   A. E. HUGHES
May 1974   D. POOLEY

Table of Physical Constants and Conversion Factors

	Symbol	Value
Planck constant	h	6.626×10^{-34} J s
Electron mass	m_e	9.109×10^{-31} kg
Proton mass	m_p	1.672×10^{-27} kg
Neutron mass	m_n	1.675×10^{-27} kg
Electron charge	e	1.602×10^{-19} C
Boltzmann constant	k	1.381×10^{-23} J K^{-1}
Speed (velocity) of light	c	2.998×10^{8} m s^{-1}
Avogadro number	N_A	6.023×10^{23} mole^{-1}
Permittivity of free space	ϵ_0	8.854×10^{-12} F m^{-1}
Permeability of free space	μ_0	1.257×10^{-6} H m^{-1}
Bohr radius	a_0	5.285×10^{-11} m
Energy associated with one electron volt	eV	1.602×10^{-19} J
Wavelength of a photon with energy one electron volt	—	1.240×10^{-6} m

Table of Symbols

a	Lattice parameter	N_e	Total number of electrons per unit volume
A_r	Relative atomic mass		
$\mathbf{b}$	Burgers vector	N_o	Number of atoms per unit volume
B	Magnetic flux density		
$\bar{c}$	Mean velocity of heat carriers	S	Entropy
		T	Temperature in kelvins
C	Heat capacity	U	Internal energy
C_V, C_p	Heat capacity at constant volumt, pressure	$U(R)$	Interaction energy of two atoms distance R apart
D	Diffusion coefficient		
$\mathscr{E}$	Electric field	$\bar{v}$	Drift velocity of an electron in an electric field
E	Energy (general)		
E_B	Binding energy, barrier height		
		V_a	Atomic volume
E_S	Schottky defect formation energy	W	Probability
		z	Projectile charge (in units of the electron charge)
E_F	Frenkel defect formation energy		
E_p	Schottky pair formation energy	Z	Atomic number
		α	Absorption coefficient
E_M	Activation energy of motion	β	Polarizability of an atom
		γ	Surface energy
E_g	Electronic band gap	ϵ	Permittivity
E_m	Maximum energy transferred in a collision	ϵ_r	Relative permittivity
		ϵ_{ro}	Relative permittivity at zero frequency
E_d	Displacement energy		
F	Helmholtz free energy	θ_D	Debye temperature
g	Force constant for atomic vibration	λ	Wavelength
		μ	Permeability
G	Shear modulus	ν	Frequency
I	Radiation intensity	ρ	Density
j	Current density	$\rho(\nu)$	Density of vibrational states
J	Particle flux		
K	Thermal conductivity	σ	Electrical conductivity; stress; cross section
$K(E)$	Differential cross-section with respect to energy E		
		$\sigma(\theta)$	Differential cross-section with respect to angle
l	Mean free path		
M_r	Reduced mass	τ	Collision time
N	Number of atoms or molecules in a solid	ψ	Wave-function

CHAPTER 1
atoms and orbitals

1.1. *Introduction*

The solid state is very familiar to all of us, but it is often surprising to realize the varied forms in which solids present themselves in every day life. On the one hand are materials such as glass and quartz which are hard, brittle, transparent and excellent insulators, and on the other, metals like copper and lead which are soft, ductile, opaque and good conductors of electricity. The explanation of all these differences lies in the way the solid is built up from atoms or molecules. In this book we are mainly concerned with the way in which defects in the atomic structure of solids affect their physical properties, or, in many cases, control them. Since one important way in which structural defects can be produced is by irradiation, the interaction of solids with radiation of all kinds is a dominant part of this theme. However, it is impossible to discuss defects without using some ideas about the properties of atoms, molecules and perfect solids, and it is therefore necessary to begin with an introduction to the basic science of the solid state. We can then build on a foundation of understanding about the differences between various classes of solids, and how this is accounted for in terms of their atomic structure.

Before being able to discuss the way in which collections of atoms are held together to form solids (and liquids), we must be quite clear about the fundamental properties of the atoms themselves. In this first short chapter we shall describe very briefly those aspects of atomic and molecular structure which we need in order to discuss properly the properties of solids. A more extended account of atomic structure and its explanation in terms of quantum theory, at a level consistent with this book, may be found in *Elementary Quantum Mechanics*. The second chapter then describes the way in which solids are built up from atoms, the explanation of the forces which hold these atoms together and the importance of the behaviour of electrons in solids in determining many solid-state properties.

1.2. *Atomic structure*

The nucleus

An atom consists of a positively charged nucleus and a number of negatively charged electrons. Atoms of different elements in the periodic

1

table have different numbers of electrons; for example, a hydrogen atom has only one electron, an iron atom 26 electrons and a uranium atom 92 electrons. The number of electrons in the atom is called the *atomic number,* denoted by Z. Since the atom as a whole is neutral, the positive charge on the nucleus must balance the total negative charge of the electrons. It follows that the charge on the nucleus is $+Ze$, where e is the magnitude of the charge on the electron. The nucleus itself consists of protons and neutrons. Each proton carries a charge $+e$ so there are Z protons in the nucleus. The neutron carries no charge. The number of neutrons in the nuclei of light elements ($Z \leq 25$) is roughly the same as the number of protons, but for a given value of Z there may be stable nuclei with different numbers of neutrons. For example, hydrogen ($Z = 1$) occurs naturally with a nucleus consisting of one proton (99·985 per cent abundant) or with a nucleus of one proton and one neutron (0·015 per cent abundant). These different forms of the same element are called different *isotopes.* The isotope of hydrogen whose nucleus contains one neutron is frequently called deuterium. Most elements have more than one stable isotope, and in addition it is possible to add extra protons or neutrons to make unstable, radioactive isotopes. The total number of protons and neutrons in the nucleus is called the *mass number* and is usually written as a superscript against the chemical symbol of an atom to distinguish different isotopes. Deuterium would thus be written ^{2}H. *The relative atomic mass A_r of an isotope is defined as the mass of an atom of the isotope on a scale on which the value of A_r for ^{12}C is exactly 12.* Since the proton and neutron are each about 2000 times as heavy as the electron, nearly all the mass of an atom resides in the nucleus. In round integers the value of A_r is the same as the mass number, and for the purpose of this book we shall not need to be more precise than this.

Although the nucleus contains nearly all the mass of the atom, it is very small. The radii of nuclei are given approximately by the formula

$$r \simeq 1.25 \times 10^{-15} \, A_r^{\frac{1}{3}} \text{ metre} \qquad (1.1)$$

from which it can be seen that for all known nuclei ($1 < A_r \leq 250$) r is between 10^{-15} and 10^{-14} m. Early estimates of the 'size' of atoms and molecules (about which we shall be more precise later) indicated radii of a few tenths of a nanometre, so that for most atomic purposes the nucleus may be regarded as a heavy point charge at the centre of the atom.

Electron states

The electrons in an atom are bound to the nucleus by the electrostatic attraction between positive and negative charges. The first really successful ideas about the properties of electrons in atoms came with Bohr's theory of the atom. This led later to the development of

2

quantum mechanics, which is able to provide an explanation for a wide variety of atomic and molecular phenomena.

In the quantum-mechanical picture, the electrons in an atom may occupy certain *stationary states* where their energy remains constant. The possible energies corresponding to stationary states are not continuous, but are 'quantized' to have only discrete values. These discrete values of the energy of electrons in atoms are called the *energy levels* of the atom. The atom may make a transition from one energy level to another by the absorption or emission of radiation, the frequency ν of the radiation being given by the quantum condition

$$h\nu = E_2 - E_1 \tag{1.2}$$

where h is the Planck constant. The sharp spectral lines observed in atomic spectra are thus a consequence of the discrete energy levels corresponding to stationary states of the atom.

The simplest atom to treat with quantum mechanics (and indeed the only one for which an exact mathematical treatment is possible) is the hydrogen atom. The possible energy levels for the single electron in a hydrogen atom are given by quantum mechanics as

$$E_n = -\frac{m_e e^4}{8\epsilon_0^2 h^2 n^2} \tag{1.3}$$

where n can be any integer from one upwards. The negative sign indicates that the electron is bound in the atom, in other words a positive energy is required to remove it. The integer n is called the

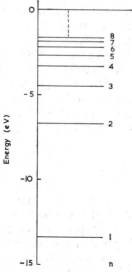

Fig. 1.1. Bound energy levels for the hydrogen atom.

3

principal quantum number. Some of the possible bound energy levels for the hydrogen atom are shown in figure 1.1. Notice that they get closer and closer together as n becomes larger. The normal state of a hydrogen atom is one in which the electron occupies the lowest possible level $n=1$. In this state an energy E_1 is required to remove the electron from the atom. This is called the *ionization energy* of the hydrogen atom and is equal to $2\cdot14\times10^{-18}$ J ($=13\cdot4$ eV, since 1 eV is approximately $1\cdot6\times10^{-19}$ J).

As well as the principal quantum number there are two other quantum numbers which are important for our discussion. The first of these relates to the angular momentum of the electron around the nucleus which, like the energy, is also quantized. The unit of angular momentum is $h/2\pi$ and the total angular momentum can have only certain multiple values of this quantity. These values are $\sqrt{[l(l+1)]}$, where the quantum number l can have integral values from zero upwards. However, there is a further restriction in that l must be less than or equal to $(n-1)$ for a given stationary state with principal quantum number n. Thus for $n=1$ we can only have $l=0$, but for $n=2$ we can have $l=0$ or $l=1$. It also turns out in the quantum-mechanical treatment that there are $(2l+1)$ distinct states all with the same value of l (corresponding to different orientations of the angular momentum of the atom). The number of these stationary states with a given value of n is thus

$$\sum_{l=0}^{n-1} (2l+1)=n^2 \tag{1.4}$$

The second additional quantum number is connected with the spin angular momentum of the electron. This is quantized to have values of $+h/4\pi$ or $-h/4\pi$ along any given axis, in other words $\pm\frac{1}{2}$ times the basic unit $h/2\pi$. This half-integral electron spin quantum number is required to explain many features of atomic structure and has its theoretical foundation in relativistic quantum theory. Since there are two possible spin quantum states for each value of n and l, the *total* number of distinct quantum states of an electron should be double that in eqn.(1.4), $2n^2$. This is a very important number, as we shall see in the next section.

1.3. *Electron shells*

The quantum mechanics of many-electron atoms becomes very complicated because, in addition to the attraction between the nucleus and each electron, the electrons repel one another through electrostatic forces. The equations of quantum mechanics then became mathematically intractable except by approximate methods of solution. However, for a nucleus with fairly large Z the mutual repulsion of two electrons will be substantially smaller than the attraction of each to the nucleus, and a valuable first approximation to the many-electron atom is to neglect the inter-electron repulsion. If this approximation is made, then every electron in the atom has available to it a set of stationary states like those in the hydrogen atom, except that the energy levels as given by eqn. (1.3) should include an extra factor Z^2 in the numerator. Stationary states of the whole atom then consist of each electron occupying a distinct state described by its quantum numbers. The way the electrons are distributed over these states is governed by the *Pauli exclusion principle*, first introduced empirically, but again having an explanation in the quantum theory. This principle states that no two electrons should occupy the same quantum state. Only $2n^2$ electrons can therefore occupy the states with principal quantum number n, since this is the number of distinct quantum states with the same n.

All the states with a given value of n comprise what is known as an *electron shell*. The shell with $n=1$ is known as the K shell, $n=2$ is the L shell, $n=3$ the M shell and so on. States with a given value of l within each shell are labelled with the letters s, p, d, f for $l=0, 1, 2, 3$. A sub-shell is labelled by n and l, for example, 1s, 2p, 3d, and there are $2(2l+1)$ states in each sub-shell. Table 1.1 shows the labelling of the electron shells and the number of available states in each.

Table 1.1. Electronic shells and sub-shells.

Shell quantum number, n	Name	Maximum number of electrons	Number of electrons in sub-shells			
			s	p	d	f
1	K	2	2	–	–	–
2	L	8	2	6	–	–
3	M	18	2	6	10	–
4	N	32	2	6	10	14

Using the concept of electron shells and the exclusion principle, the electronic structure of all the elements can be built up. Helium has two electrons, so both go in the K shell. Lithium has three electrons, so that two go in the K shell and one in the 2s sub-shell. Sodium has 11 electrons: two in the K shell, eight in the L shell leaving one to go in the 3s sub-shell. The chemical properties of the elements are determined mainly by the electrons in the highest shell, so the chemical similarity of lithium and sodium arises because both have a single electron in an s sub-shell. In this way the groups of elements within the periodic table follow from their electronic shell structure. Shells or sub-shells which are filled to capacity with electrons are generally very stable and chemically inert. The inert gases such as helium and neon, for example, have completely full shells. In a related way, the valency of an element depends on its shell structure. Sodium is monovalent because it has one electron outside a set of full shells. Magnesium has one more electron

5

than sodium and so is divalent. Elements with multiple valency have several electrons outside full shells and can share these in more than one way with another element in forming a compound. For example, iron has filled K and L shells and filled 3s and 3p sub-shells. The remaining eight electrons are distributed so that six are in the 3d sub-shell and two in the 4s sub-shell. As a result of having these eight electrons outside the stable shell structure, iron can be divalent or trivalent, as is well known from its ability to form the three oxides FeO, Fe_2O_3 and Fe_3O_4.

1.4. *Orbitals*

In our discussion of quantum numbers and shell structure, we have not mentioned anything about the spatial distribution of electrons in an atom. Although in quantum mechanics the energy of an electron in a hydrogen atom has discrete values, its position relative to the nucleus does not. Indeed, all that can be deduced about the position of the electron in a particular stationary state is the *probability* of finding it in a certain region of the atom. This probability is given by the squared modulus $|\psi|^2$ of a spatial function ψ called the *wave-function* of the electron, and quantum mechanics tells us the form of the wave-function belonging to a particular quantum state. The form of the wave-function is also known as the *orbital* of the electron, since it shows how the negative charge of the electron 'orbiting' in the atom is distributed statistically in space. The probability of finding the electron in a volume element dV is then given by $|\psi|^2\,dV$, so *on average* the charge

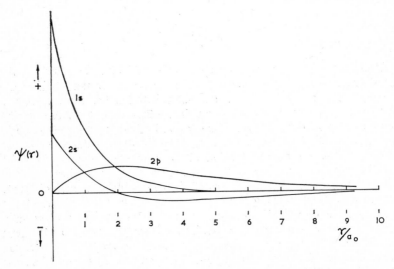

Fig. 1.2. Radial dependence of wave-functions for the hydrogen atom.

6

density in this region will be $e|\psi|^2$. Figure 1.2 shows the radial distribution of the 1s, 2s and 2p orbitals in the hydrogen atom. The 1s orbital has the form

$$\psi = \text{constant} \times \exp\left(-r/a_0\right) \tag{1.5}$$

so that the charge density falls off exponentially with increasing distance from the nucleus. The quantity a_0 is known as the first Bohr radius and has the value

$$a_0 = \frac{\epsilon_0 h^2}{\pi m_e e^2} = 0\cdot053 \text{ nm} \tag{1.6}$$

This gives us our first theoretical indication of the concept of the radius of an atom. A hydrogen atom in the 1s state (that is, with its electron in the 1s orbital) has a charge distribution which extends over a distance of about a_0, so this is a rough definition of the radius of the atom.

In atoms with larger Z one would expect the orbitals to become more compact because the electrons are attracted more strongly to the nucleus. In fact the full expression for the value of a_0 in the 1s orbital (eqn. (1.5)) has Z in the denominator so that, for example, a_0 becomes $0\cdot026$ nm for helium and $0\cdot018$ nm for lithium. However, figure 1.2 shows that the charge distribution in orbitals of higher n is greater in radial extent than those of small n. In an atom with a filled 1s shell, the two electrons in this shell will therefore shield to some extent the other electrons in higher shells from the nuclear charge. This is, in effect a crude way of allowing for the inter-electron repulsion which we ignored earlier. As a result of this shielding effect, the electrostatic attraction of the nucleus towards electrons in unfilled shells is reduced to a size appropriate to a small effective value of Z. Also, we must remember that as Z increases electrons must be placed in higher shells whose orbitals are in any case more extended. Both these (related) effects combine to ensure that the radial distribution of charge in atoms does not change markedly throughout the periodic table. This may be contrasted with the radii of nuclei, which increase monotonically with increasing mass number (see eqn. (1.1)).

To summarize the discussion so far, electrons in atoms are distributed over the electron shells according to the Pauli exclusion principle. The chemical properties of atoms are determined by the outermost electrons in the atom, which are those in the highest shell. The radial distribution of electronic charge in the atom is given by the spatial extent of the orbitals, and extends typically over a distance of a few tenths of a nanometre from the central nucleus.

1.5. Interactions between atoms

There must be some attraction between atoms since we know they form molecules and solids. We shall leave until the next chapter a detailed discussion of these attractive forces, but some general features are worth mentioning here since they have a bearing on the concept of atomic radii. If two atoms which have a chemical affinity for one another are separated by a distance greater than the separation of the two atoms in the molecule, there will be an attractive force between them. If there were not, there would be no tendency for them to combine into a molecule. Suppose now the value of the inter-atomic separation R is reduced. Once the atoms get closer together than 0·5 nm or so, there will be some overlap of the orbitals corresponding to the filled electron shells of the two atoms, and some repulsive force between the two sets of electron charge distributions. Also, there will be some repulsion between the two nuclei since they will not be completely screened from each other by the electrons. The result will be a repulsive force between the two atoms once they approach each other closely enough. The equilibrium separation R_0 of the atoms in the molecule will be determined by a balance between the attractive force and this repulsive force, as shown in figure 1.3 (a). It is often more convenient to discuss the problem in terms of an energy of interaction $U(R)$ rather than forces, the force being $-dU(R)/dR$. The repulsive force may thus be derived from a repulsive energy, and an energy diagram constructed as in figure 1.3 (b). The equilibrium position R_0 is where the total force is zero, that is $dU/dR = 0$, and the energy is a minimum.

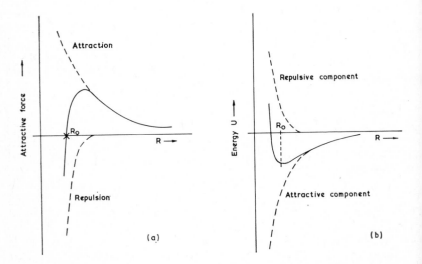

Fig. 1.3. The equilibrium of two atoms under the combined action of attractive and repulsive forces (a) and their corresponding potential energies (b).

8

The repulsive energy increases sharply once the charge distributions overlap, so that the equilibrium separation of the atoms in the molecule is close to the point at which significant overlap occurs. It is found in practice that it is possible to define an atomic radius for an element in terms of the separation of its atoms in molecules or solids, deduced radii usually being in the range 0·1–0·2 nm. From what we have said about repulsive forces it is clear that this is consistent with our previous supposition of an atomic size based on the radial extent of orbitals. There are thus sound reasons for thinking of atoms as small spheres with well-defined radii, a picture which is useful in discussing many of the properties of solids. The combination of attractive and repulsive forces as illustrated by figure 1.3 is also extremely valuable in discussing the bonding of atoms in solids, as we shall see in the next chapter.

CHAPTER 2

the structure and properties of perfect solids

2.1. *The structure of solids*

All solids are made from atoms packed closely together in a condensed phase. The density of a solid does not change much when it melts, so the number of atoms per unit volume must be similar for solids and liquids. As we have seen, the radius r of a typical atom is about 0·15 nm, so in a solid or liquid there are about $(2r)^{-3} \sim 4 \times 10^{28}$ atoms m^{-3}. What is then the difference between a solid and a liquid, and what is responsible for the rigidity of solids as opposed to the free flow of liquids? In most simple compounds the answer is straightforward: in the solid state the atoms are organized into a regular ordered crystal lattice of repeated *unit cells*, and are constrained in a particular site by the close packing of their neighbours. In a liquid there is no long-range lattice structure, and the atoms jostle one another around as they move fairly freely through the whole volume. The essential difference between a simple crystalline solid and a liquid is indicated by figure 2.1.

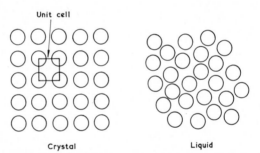

Unit cell

Crystal   Liquid

Fig. 2.1.   The atomic structure of a crystalline solid and a liquid.

The fact that most solids have a regular crystal structure is familiar to us from the regular shapes that are commonly found in natural minerals. Even grains of common salt can be seen to have a cubic shape. These are the macroscopic manifestations of the crystal structure, and figure 2.2 shows some of the atomic arrays forming the crystal lattice of a number of simple solids with cubic crystal structure. Many more complicated structures are also possible, with lower symmetry than the cube.

Having ascribed the difference between liquids and solids to ordering of atoms in the solid phase, we must extend this simple picture a little.

10

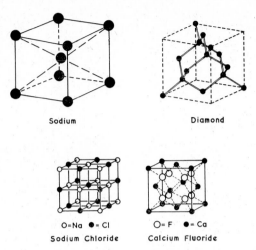

Sodium

Diamond

O=Na ●= Cl   O= F ●= Ca
Sodium Chloride   Calcium Fluoride

Fig. 2.2. Some common crystal structures.

Quite complicated chemical compounds form crystals: emeralds, for example, are crystals of beryl, $Be_3Al_2Si_6O_{18}$, with some of the aluminium ions replaced by chromium impurities. However, some very simple compounds such as silica (SiO_2) may enter a solid phase without having a regular crystal structure. It is almost as if a liquid had become so viscous that the atoms are effectively stuck in one place, even though the structure is more or less random and liquid-like. Such amorphous or vitreous solids are quite common, glass being the most familiar example. Plastics also have irregular structures, but in this case they consist of entanglements of very long molecules.

The distinction between crystalline and amorphous solids is a very clear one on a microscopic scale. A less fundamental point, but one which is very important in practice, must be considered in discussing the macroscopic structure of crystalline solids. It is seldom that a sample of a crystalline solid is actually a single crystal, that is, a continuous array of atoms ordered over the whole volume of the material. Imagine a mass of molten salt solidifying at its melting temperature. Different parts will begin to solidify independently with no long-range coherence of the crystal structure, and the result will be a solid of small inter-locked individual crystals. Such a solid is known as a *polycrystal* and the individual crystals as crystallites or as *grains*. The size of the grains can vary widely depending on the mode of solidification and subsequent treatment, and many of the properties of the material such as the mechanical strength are critically dependent on the grain size. The regions between the grains, where the different crystallites do not fit together properly, are known as *grain boundaries,* see figure 2.3. Many of the materials in everyday life, such as most metals, pottery and brick,

11

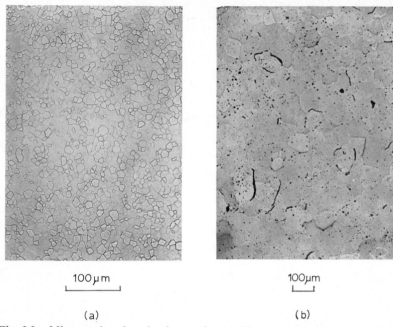

100μm

(a)

100μm

(b)

Fig. 2.3. Micrographs of grains in a polycrystalline material. (*a*) shows a fine-grained polycrystal of MgO in which full contact is achieved between all the grains; (*b*) shows a coarse-grained polycrystal of MgO in which there are pores (the dark areas) at the grain boundaries and within some of the grains. The grain boundaries are made visible by etching the sample in orthophosphoric acid (courtesy of R. A. J. Sambell).

are polycrystalline, and as a result their macroscopic properties appear very different from those of single crystals of the constituent compounds.

Since solids take on many forms, it is impossible to give a complete account of defects and radiation effects in all situations. Only in a few of the simpler structures has a genuine understanding of these effects been achieved so far. In this book our concern will be with fairly simple materials which solidify in a crystalline state, often as poly-crystals. This nevertheless includes a wide class of solids with important technological applications in reactors, spacecraft and other radiation environments, as well as materials where defects and radiation effects are used to good effect.

2.2. *Crystal bonding*

So far we have discussed the general properties of solids from a structural point of view. Let us now look a little more closely at the behaviour of solids on an atomic scale. The first question to answer is the very basic one of what holds a solid together. In a gas the atoms or molecules move more or less independently of one another, but in a

12

crystalline solid they are locked into a regular array. In all solids the atomic forces which preserve the array arise from the electrostatic forces between the electrons and nuclei of the atoms in the solid, but it is useful to classify distinctive types of bonding according to the way these forces manifest themselves. These classifications do not represent hard and fast divisions. They are simply convenient ways of distinguishing between different types of behaviour of the electrons in the solid.

Ionic bonding

Some simple compounds such as salt (NaCl) are held together because the sodium and chlorine atoms in the solid are themselves oppositely charged (they are then strictly called ions rather than atoms). To see the nature of this type of bond, it is simplest to consider first a single molecule of NaCl isolated in the vapour phase. Of the 11 electrons in the sodium atom, 10 occupy states which form full electron shells. These filled shells form a highly stable structure, since these ten electrons on the sodium atom occupy the same shells as do all ten electrons of the chemically inert neon atom. The eleventh electron in the sodium atom has to go into a higher shell, and occupies a 3s orbital. In this orbital the electron is further from the nucleus than those in the filled shells and the attractive electrostatic field of the nucleus is effectively shielded by the ten electrons in the closed shells. This eleventh electron is thus quite weakly bound to the atom: it takes only $5\cdot14$ eV of energy to remove it. This energy is the first ionization energy of the sodium atom. The sodium atom (Na) may thus be regarded as a highly stable (and inert) ionic core (Na^+) with one rather loosely bound electron in a 3s orbital.

The chlorine atom has 17 electrons, which is enough to fill the 1s, 2s, 2p and 3s sub-shells (12 electrons altogether), but one short of filling the 3p sub-shell. An eighteenth electron can join the incomplete 3p sub-shell and is not completely shielded from the electrostatic attraction of the nucleus by the other electrons, although the net charge of the atom it is joining is zero. This is because all the electrons in the 3p orbitals are about equally distant from the nucleus and none can be completely shielded from it by the others. Thus the chlorine atom can attract an electron to complete the 3p sub-shell and make an ionic core (Cl^-) which has the same stable shell structure of electrons as the argon atom. The energy gain in this process, known as the electron affinity of chlorine, is $3\cdot71$ eV.

The total amount of energy required to take the 3s electron from a sodium atom and put it in a 3p orbital of the chlorine atom is thus $5\cdot14-3\cdot71=1\cdot43$ eV. This is still positive, so why should this transfer happen in a molecule or crystal of NaCl? If the Na^+ and Cl^- ions stayed very far apart, their combined energy would indeed be greater than that of a Na atom and a Cl atom. However, the two ions are

13

oppositely charged, and attract each other with an electrostatic force $e^2/4\pi\epsilon_0 R^2$, where R is their distance of separation. When they are a distance R apart, their mutual potential energy is $-e^2/4\pi\epsilon_0 R$. Thus if they are able to get close enough to each other, this negative energy can outweigh the small positive energy (1·43 eV) required to form the ions. The total energy is then less (that is, more negative) than that of the two separated atoms, and it will take a finite force to separate the two ions. They are therefore *bonded* together in an electrostatic or *ionic* bond. The scheme is illustrated in figure 2.4.

$$\text{Total energy} = 5\cdot14 - 3\cdot71 - \frac{e^2}{4\pi\epsilon_0 R}$$

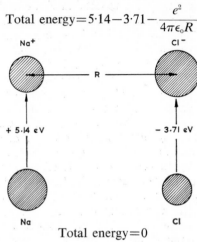

Fig. 2.4. The origin of the cohesive energy of NaCl.

How closely can the two ions approach each other, or in other words, how strong is the bond? The equilibrium separation of the two ions is determined by a balance between the attractive electrostatic force between the ion cores and the repulsive forces as the electron shells begin to overlap, just as was shown in figure 1.3. The radii of the Na$^+$ and Cl$^-$ ions are 0·098 nm and 0·181 nm respectively, so that the equilibrium position occurs when R is close to the sum of the two ionic radii, $R_0 \simeq 0\cdot3$ nm. The bonding energy E_B is then approximately

$$E_B \simeq \frac{e^2}{4\pi\epsilon_0 R_0} \simeq 8 \times 10^{-19} \text{J} = 5 \text{ eV for } R_0 = 0\cdot3 \text{ nm}$$

(The repulsive potential energy reduces the actual value of E_B by about 10 per cent from that given above.) The ionic bond is thus quite strong, as might be expected from the stability of molecules such as NaCl.

A crystal of NaCl is held together by these same electrostatic forces as occur in a molecule of NaCl, the actual crystal structure being that which provides the minimum total potential energy (and therefore maximum binding energy). The crystal structure of NaCl is shown in figure 2.2. Each Cl$^-$ ion is surrounded by six Na$^+$ ions in the closest

14

sites (nearest neighbours), 12 other Cl⁻ as next nearest neighbours, and so on. The total electrostatic potential energy is found by calculating all the interactions between pairs of ions and (after a rather complicated sum) is found to be

$$E = -\frac{1 \cdot 75 \; Ne^2}{4\pi\epsilon_o R}$$

where N is the total number of NaCl molecules in the crystal and R the separation of the nearest Na⁺ and Cl⁻ ions. The factor $1 \cdot 75$ is called the *Madelung constant* for the NaCl crystal structure. The equilibrium value of R is found by minimizing the total potential energy, just as in figure 1.3 (*b*), and the binding energy per NaCl molecule in the crystal is found to be $12 \cdot 6 \times 10^{-19} J = 7 \cdot 9$ eV.

Crystalline solids which are held together by electrostatic forces between oppositely charged ions are known as *ionic* or *polar* crystals. Examples of ionic crystals are the alkali halides such as NaCl, KCl and LiF, alkaline earth fluorides such as CaF_2, and most oxides. For ions of charge $\pm ze$ the ionic bond energy per molecule is, by analogy with our previous discussion of NaCl, of order

$$E_B \simeq \frac{z^2 e^2}{4\pi\epsilon_o R_o}$$

The high melting temperatures and physical strength of oxides, so important in making them technically useful materials, stem from the fact that the oxygen atom has four electrons in the 2p sub-shell. Two electrons are therefore required to fill it, and oxygen is usually found in the divalent state as an O^{2-} ion with $z=2$. An atom like Mg with two outer 3s electrons is also divalent, losing these to become Mg^{2+}. A crystal of MgO, which has the NaCl structure, thus has a binding energy roughly four times that of an alkali halide. This stronger bonding is illustrated by its melting temperature of 2800 °C, which compares with 800 °C for NaCl.

Covalent bonding

The ionic bond is an example of complete exchange of electrons from one type of atom to another. The *covalent* (or *homopolar*) bond, on the other hand, is a result of the sharing of a number of electrons between similar atoms. The simplest covalent bond is that which keeps two hydrogen atoms together to form a molecule. Each hydrogen atom has one electron, which in a free atom occupies a 1s orbital in an unfilled shell. As two hydrogen atoms approach each other closely enough for the two 1s orbitals to overlap, the two electrons experience not only the attractive force of their own nuclei, but also the attractive force of the nucleus belonging to the other atom. In the molecule the two electrons are shared equally between the two nuclei, so that each electron gets the benefit of the attractive force between itself and two positive charges. In a single hydrogen atom each electron only interacts with one nucleus, so the interaction with both nuclei represents an attractive force between the two hydrogen atoms. The explanation of the covalent bond in terms of quantum mechanics is one of the foundations of molecular chemistry, and is discussed in most text-books on the subject. An important difference between a covalent bond and an ionic bond is that the number of atoms which can participate in the former is strictly limited. For example, there is no significant attraction between a hydrogen molecule and a third hydrogen atom. An ion in an ionic crystal, on the other hand, has a positive attraction to any number of ions of the opposite charge. The covalent bond is said to be saturated, and the ionic bond unsaturated. The reason for the exclusive nature of the covalent bond is the Pauli exclusion principle. In the H_2 molecule the 1s orbitals of each atom combine to form one molecular orbital which corresponds to the state of lowest energy (and thus strongest bonding). Only two electrons may occupy this *bonding* orbital. The electron of a third hydrogen atom cannot therefore enter this low energy state, but must go into an orbital of higher energy. This is known as an *antibonding* orbital, because when it is occupied there is repulsion rather than attraction between the atoms. The hydrogen molecule does not therefore bind a third atom to form H_3.

The covalent bond can be very strong like the ionic bond, and is responsible for the atomic forces in elemental solids such as diamond, silicon (Si) and germanium (Ge). The hardness of diamond demonstrates the strength of the covalent bond, in this case the sharing of the four outer electrons (two in the 2s orbital, two in the 2p orbital) on each carbon atom, giving a bond energy of 7·3 eV. The fact that there are four shared electrons per atom is responsible for the crystal structure of diamond, in which each atom is surrounded by four others in a tetrahedral arrangement, figure 2.2. The bonding orbitals between a carbon atom and its four neighbours have the form shown schematically in figure 2.5. Two electrons occupy each bonding orbital as indicated.

Many crystals composed of different elements are held together by a mixture of covalent and ionic bonds. For example, gallium arsenide (GaAs) is made up of Ga atoms with three electrons outside completely full electronic shells, and As atoms with five outer electrons. The bonding is probably rather like that in diamond, and indeed the crystal structure is very similar, each Ga atom being surrounded by four As atoms in a tetrahedron, and vice versa. The sharing of four electrons on each atom with its neighbours would, however, mean that one electron is transferred from the As to the Ga atom. This would give an ionic component to the bonding of GaAs. Conversely, most oxides are not entirely ionic, and have some covalent bonding which makes z less than two. This emphasizes that the bonding classifications are convenient rather than rigorous.

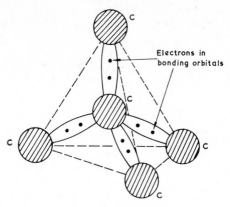

Fig. 2.5. Illustration of tetrahedral bonds in diamond.

Metallic bonding

The bonding of metals may be regarded as an extended covalent bond, in which the shared electrons are fairly free to wander throughout the whole crystal. In this sense they form a uniform negatively charged glue spread throughout the material, which holds the remaining positively-charged ions (the closed shells) together by electrostatic attraction. Despite the non-localized nature of the metallic bond it is fairly strong; if it were not, then metals would be far less useful in their varied applications.

Van der Waals bonding

Both ionic and covalent bonds require the presence of loosely bound outer electrons (often called *valence electrons*) on the interacting atoms. This is not the case for chemically inert atoms which only have closed electronic shells, such as neon and argon. These do form solids at low enough temperatures, so there must be some kind of attractive interaction even between uncharged atoms with closed shells. The low melting temperatures, however, suggest that these forces are relatively weak. They are known as *van der Waals* or *London* forces. The way they occur may be seen with the following very simplified argument. The negative electrons and positive nucleus on one atom give rise to a rotating electric dipole moment as the electrons rotate around the nucleus. Of course, these dipole moments average to zero over a finite time, because the electron orbitals are symmetric about the nucleus. However, the momentary dipole moment on one atom can induce a dipole moment on a second atom by polarizing its electron orbitals, and the two dipoles then attract each other. Even though the individual dipole moments average to zero, the forces at any moment in time are attractive, so there is a permanent attractive interaction between

17

the atoms. The way in which the van der Waals potential energy varies with the atomic separation R may be seen as follows. The electric field $\mathscr{E}$ due to a dipole moment p_1 on one atom is

$$\mathscr{E} \propto \frac{p_1}{R^3}$$

This field induces a dipole moment $p_2 = \beta\mathscr{E}$ on the second atom, where β is its polarizability. The potential energy of this second dipole in the field $\mathscr{E}$ is $-\frac{1}{2}\beta\mathscr{E}^2$, so that the van der Waals energy is proportional to $-\beta p_1^2/R^6$ [compare with the electrostatic potential energy of charged ions which varies as $1/R$]. The inverse sixth power may also be derived more rigorously using quantum mechanics, and shows that the van der Waals interaction has a relatively short range. It is also fairly weak, with a binding energy E_B of usually only a fraction of an electron volt. Heavy atoms with large polarizabilities (because they have a large number of electrons) have the largest van der Waals interactions. The weakness is reflected in the low melting temperatures of inert gas solids (in fact liquid helium does not solidify under atmospheric pressure, even at absolute zero!).

Van der Waals forces contribute a component to the bonding of all solids, but are often negligible in the face of the stronger ionic and covalent bonds. As we have seen, they are important for rare-gas solids, and are also responsible for the bonding in many organic crystals.

Hydrogen bonds

Finally, some crystals and molecules are held together by what is called a *hydrogen bond*. In an extreme case this can be thought of as a bond in which a hydrogen atom loses its electron to become a bare proton. The positive charge of the proton may then bind two negative ions like F^- and O^{2-} together. This type of bonding is very common in organic molecules and plays an important part in the cohesive forces between water molecules. It is seldom important in simple inorganic crystals. Like all the other classifications of bonding, its origins lie in electrostatic forces, in this case involving the electrons of the ions and the proton they share.

Table 2.1 summarizes the different types of bonding we have described and gives some typical examples.

Table 2.1. Classification of the bonding of solids.

Class	Typical bond strength eV per atom	Examples
Ionic	$8z^2$	NaCl, MgO, CaF_2
Covalent	4–8	Diamond, Si, Ge
Metallic	1·5	Na, Al
Van der Waals	<1·0	Ar, anthracene
Hydrogen bond	0·1	H_2O

2.3. *The electronic structure of solids*

Electrons in solids

The next basic question concerning the nature of solids is the distinction between metals, semiconductors and insulators. The resistivity of a metal like copper at room temperature is about 10^{-8} Ωm, whereas a good insulator may have a resistivity as high as 10^{15} Ωm. Semiconductors have resistivities in the range 10^{-4}–10^7 Ωm. In special cases the resistivity may have even lower or higher values than those quoted here. These vast differences in the electrical behaviour of different solids originate in the way electrons behave in crystal lattices. So far we have regarded the electrons in the solid as largely belonging to a particular atom (except possibly in the case of metals where our brief discussion of metallic bonding in Section 2.2 suggested that at least some of the electrons were able to wander through the whole solid). For example, in discussing ionic bonding we talked about individual ions with electrons bound in closed electronic shells, and in talking about the covalent bond we visualized electrons shared between pairs of atoms. This is an accurate description of molecules and a convenient picture for discussing bonding in solids, since there one is interested in the relatively large energies required to separate the solid into its constituent atoms. In discussing the electrical properties of solids, one must look in rather finer detail at the factors which determine the energy levels of the electrons, and recognize that there is no justification for associating a particular electron with a particular atom. Not only are all the electrons indistinguishable, but any one of them is just as likely to be found on one atom as another. The electrons must therefore be regarded as belonging to the solid as a whole, and not to any one particular atom. This concept is no more unusual than the more easily accepted idea that electrons in molecules belong to the molecule rather than its constituent atoms. We have already used this point of view in Section 2.2 when we discussed the hydrogen molecule.

So we arrive at a picture in which the electrons in a solid are shared between all the atoms and in a sense are travelling through the whole crystal. In classical mechanics this gives rise to a complicated situation, but fortunately electrons obey the rules of quantum mechanics, which are simpler to the extent that it is not permitted to ask questions about the movements of a particular electron, but only about the energy of the electrons and the probability of finding an electron in a particular position. Those properties of crystals which are controlled by the motion of the electrons therefore depend on the quantum states of electrons moving through the solid, just as the properties of an atom depend on the atomic orbitals occupied by its electrons. We must therefore now discuss these crystal quantum states, and in particular the energy levels of electrons in solids.

19

Energy bands

The energy levels of electrons in an isolated atom are discrete, each level separated from the next by a finite energy often of order ~ 1 eV (see figure 1.1). The electrons moving in a crystal are not localized on any one atom, and experience an electrostatic potential which is periodic such as in figure 2.6. Quantum mechanics then shows that the energy

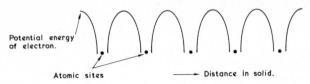

Potential energy of electron.

Atomic sites   ⟶ Distance in solid.

Fig. 2.6. The periodic electrostatic potential energy of an electron moving in a crystalline solid.

levels of the electrons in the crystal fall into well-defined energy *bands* rather than discrete levels as in atoms, and the wave-functions of the electrons extend throughout the whole crystal. The simplest view of the formation of these energy bands is to imagine the atoms of a solid being brought together from a large separation to form the crystal. If only two atoms were brought together to form a molecule, each atomic energy level would split into two, as in figure 2.7 (*a*). We have already

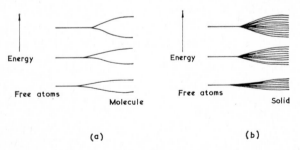

Energy   Energy

Free atoms   Free atoms

Molecule   Solid

(a)   (b)

Fig. 2.7. Energy levels as a function of inter-atomic separation in (*a*) a molecule, (*b*) a solid.

come across this in the case of the hydrogen molecule in Section 2.2: the 1s levels on each H atom split up to make one bonding orbital and one antibonding orbital. If N atoms are brought together, each single atomic level splits into N sub-levels which together form a *band* of energy levels, see figure 2.7 (*b*). Since in a solid $N \simeq 4 \times 10^{28}$ m^{-3}, the sub-levels are so close together that the possible energies within each band can be regarded as more or less continuous. This is the essence of the *band theory* of solids.

20

The electronic properties of the solid are determined by the number of electrons which can occupy the sub-levels in the energy bands. The Pauli exclusion principle tells us that only two electrons, of oppositely directed spin angular momentum, may occupy a sub-level in an energy band. In our example above, each band may therefore contain $2N$ electrons. A more careful approach to the problem shows that N should be the number of *unit cells* in the crystal rather than the number of atoms. This is because if there is more than one atom in a unit cell of the crystal each atomic energy level is responsible for more than one band.

What does the band theory of solids tell us about the electrical conductivity of solids? The wave-functions of electrons in energy bands have the form of waves travelling through the solid. In any band there will be as many states corresponding to waves travelling in one direction as in the opposite direction. If a current is to flow when an electric field is applied to the crystal, there must be more electrons moving in one direction than another. This is only possible if the distribution of electrons over the sub-levels in the energy band can be arranged so that more occupy (on average) states corresponding to waves travelling in the direction of electron current flow than against it. In other words, there must be empty levels in the band so as to permit this readjustment. This cannot possibly occur if the band is completely full, unless the electric field is so high that electrons are excited into a higher energy band. A solid with completely full bands thus cannot give a current flow and is an insulator: the fact that every sub-level is occupied does not allow a net movement of electrons. Conversely, metals have only partially filled energy bands. The highest energy bands in a metal and an insulator are shown schematically in figure 2.8. A good insulator has a *band gap* E_g between the highest energy

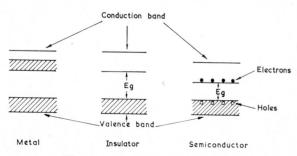

Fig. 2.8. Electronic energy bands.

filled band (called the *valence* band) and the lowest unfilled band (the *conduction* band) of several electron volts. This ensures that even at moderately high temperatures (say several hundred K) virtually no electrons are excited thermally into the conduction band. The average number of electrons which are thermally excited across a band gap E_g is roughly proportional to $\exp(-E_g/2kT)$, where k is the Boltzmann constant. At $T=300$ K, $kT \simeq 1/40$ eV, so that if $E_g=5$ eV the exponent is -100 and the number of excited electrons negligible. In a pure semiconductor such as silicon or germanium, E_g is not as large ($1\cdot14$ eV for Si, $0\cdot67$ eV for Ge) and a few electrons may be thermally excited into the conduction band. At $T=300$ K the number of conduction electrons is $2\cdot5\times10^{19}$ m^{-3} in Ge and $1\cdot4\times10^{16}$ m^{-3} in Si. The remaining partly filled valence band also allows conduction, since there are unoccupied sub-levels in the band. Electrons may then be excited in and out of these *holes* and a net current flow may exist in an applied electric field. If an electron moves into a hole, then this is equivalent to motion of the hole to the place formerly occupied by the electron, and conduction by the valence band electrons is normally viewed as the movement of holes, which are effectively positively charged. At 300 K the resistivity of *pure* Ge would be $0\cdot43$ Ωm, and that of Si $2\cdot6\times10^3$ Ωm. The semiconductor is thus a special case of an insulator, as is also shown in figure 2.8.

With the band picture it is easy to see why solids such as Na, K and Rb are metals, since they have an odd number of electrons per unit cell and therefore possess a partially filled band. Similarly, Si and Ge are expected to have a full valence band, since there are an even number of electrons per unit cell. The same applies to insulators such as the alkali halides. The situation is not as straightforward for all solids, since sometimes different bands can overlap and give a metal even when there are an even number of electrons per unit cell (for example in Ca and Zn). Nevertheless, the band theory of solids provides the physical basis for understanding the behaviour of electrons in solids, and tells us the conditions under which a solid conducts or does not conduct electricity.

Finally, a word about bands of lower energy than the valence band. In all solids there exists a series of completely filled bands below the valence band, which correspond to the filled inner electronic shells of the atoms or ions. As a general rule these bands become narrower as their energy decreases (which corresponds to the electrons being more tightly bound) and the wave-functions of the electrons when near a particular atom resemble more and more closely the atomic orbitals corresponding to a given inner shell of that atom. Properties which depend on the tightly bound inner shell electrons, such as X-ray spectra, are therefore not greatly changed in the solid relative to the free atoms. If the atoms or ions in the solid have electron shells that are all completely full, then even the valence band wave-functions may be very similar to atomic orbitals. This is the case for many simple ionic and van der Waals crystals, and explains why we could discuss these types of bonding in terms of individual atoms, despite the non-localized

22

nature of true electron states. We thus have the situation that some electrons in solids do behave as if they still belong to individual atoms, even if those responsible for any electrical conduction certainly do not.

2.4. *Vibrations in solids*

So far we have implicitly assumed that the atoms in a crystal are firmly confined to their lattice sites. The lattice site corresponds to the minimum in a potential energy curve such as that of figure 1.3 (*b*). However, the atoms may vibrate around these lattice sites, and since close to the minimum the potential well is parabolic $(U(R)-U(R_o)\simeq\frac{1}{2}g\Delta R^2$; g being a 'force constant') these vibrations are like those of a harmonic oscillator. These *lattice vibrations* are important because their kinetic and potential energy are responsible for most of the thermal energy of a solid: that is, that part of the energy of the solid which depends on its temperature. In classical terms, the higher the temperature of the solid, the larger is the amplitude of the oscillations and the more energy stored in them.

It was recognized very early on in the study of solids that the energy of the atomic oscillations was related to the heat capacity. In classical statistical mechanics the mean energy of a one-dimensional harmonic oscillator at temperature T is kT ($\frac{1}{2}kT$ is the mean potential energy and $\frac{1}{2}kT$ the mean kinetic energy, making kT altogether). Each atom in a solid may vibrate in any of three perpendicular directions: it has three *vibrational degrees of freedom*. Thus a solid of N atoms has a thermal energy of $E=3NkT$. In order to raise the temperature by 1 K it is therefore necessary to provide an energy of $3Nk$ so that the heat capacity $C_V=(\partial E/\partial T)_V$ is $3Nk$. This result holds well for most solids at fairly high temperatures ($T\gtrsim200$ K). At low temperatures C_V falls towards the value zero at $T=0$ K, as shown in figure 2.9, and this

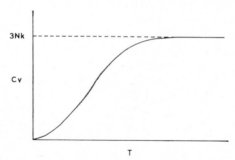

Fig. 2.9. Typical form of the heat capacity of a solid as a function of absolute temperature.

cannot be explained by the classical theory. The explanation lies in the quantized nature of the energy of the lattice vibrations. In quantum

23

mechanics the energy levels of a harmonic oscillator of frequency ν are given by

$$E_n=(n+\tfrac{1}{2})\ h\nu \qquad (2.1)$$

where n is an integral quantum number ($n=0,1,2\ .\ .\ .$). The average energy of an oscillator is then no longer kT, but is given by†

$$\bar{E}=\frac{h\nu}{\exp\,(h\nu/kT)-1}+\frac{h\nu}{2} \qquad (2.2)$$

At high temperatures $\exp\,(h\nu/kT)\approx1+h\nu/kT+\tfrac{1}{2}(h\nu/kT)^2$ and $\bar{E}\simeq kT$ as in the classical model. As $T\to0$, $\bar{E}\to h\nu/2$ and becomes independent of temperature. $h\nu/2$ is called the *zero-point energy* of the oscillator.

The central problem in the quantum theory of the heat capacity is to know the frequency (or frequencies) of the atomic vibrations. In the classical theory this is conveniently unnecessary. The simplest quantum model is the Einstein model, which assumes that each atom vibrates independently with the same frequency ν. The heat capacity then becomes, by differentiating $E=3N\bar{E}$,

$$C_V=3Nk\left(\frac{h\nu}{kT}\right)^2\frac{\exp\,(h\nu/kT)}{[\exp\,(h\nu/kT)-1]^2} \qquad (2.3)$$

This expression explains the general shape of curves such as that in figure 2.9, the value of ν required to fit experimental curves usually being about 10^{13} Hz. The agreement is not perfect, however, because experimentally C_V is proportional to T^3 at low temperatures, whereas the Einstein model predicts an exponential temperature dependence. Nevertheless, the quantization of the energy levels of an oscillator does explain why C_V falls to zero, which was Einstein's main purpose.

The inadequacy of the Einstein model is its assumption of a single vibrational frequency for all the atoms in the solid. This is a great over-simplification, since the atoms are all coupled together by the interatomic forces and do not vibrate independently. Instead a large number of coupled modes of oscillation are possible, whose frequencies are found to span a wide range, from zero up to some maximum determined by the strength of the atomic forces. There is thus a more or less continuous spectrum of frequencies ν, which may be described by a density-of-states function $\rho(\nu)$. This is defined by letting $\rho(\nu)\Delta\nu$ be the number of independent modes of oscillation with frequencies in the range ν to $\nu+\Delta\nu$, $\Delta\nu$ being a small frequency interval.

---

†The probability that the oscillator has energy E_n is proportional to $\exp\,(-E_n/kT)$. The average energy is thus

$$\bar{\bar{E}}=\frac{\Sigma_n E_n\exp\,(-E_n/kT)}{\Sigma_n\exp\,(-E_n/kT)}$$

which gives the expression in eqn. (2.2).

The total number of modes possible is then $\int \rho(\nu)\,d\nu$ and is equal to the total number of degrees of freedom of the atoms in the solid, $3N$. This is actually a general theorem of the modes of oscillation of coupled systems, which is beyond the scope of our discussion.

The density-of-states function is usually quite complex, an example being the $\rho(\nu)$ calculated for NaCl shown in figure 2.10. Some features,

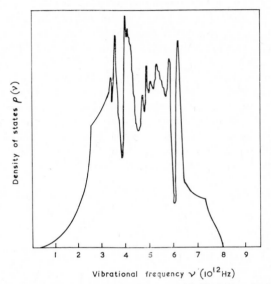

Fig. 2.10. The vibrational density of states of NaCl (after A. M. Karo and J. R. Hardy, *Physical Review*, **141**, 696 (1966)).

however, are quite general. There is always a maximum frequency ν_{max}, which is highest for strongly bonded solids with light atoms such as diamond (remember that for a harmonic oscillator $\nu^2 \propto g/M$, where g is the force constant and M the mass of the oscillator). At very low frequencies the modes of oscillation are nothing more than sound waves, with a wavelength much greater than the spacing of the atoms in the crystal. The density-of-states $\rho(\nu)$ always starts from zero and is initially proportional to ν^2. In fact all the modes can be represented by waves, but only those of low frequency travel with the velocity of sound. At higher frequencies the wave velocity depends on the detailed atomic structure of the solid.

Each mode of oscillation has an average energy given by eqn. (2.2), so that the total thermal energy of the solid is given by

$$E = \int_0^{\nu_{max}} \rho(\nu)\overline{E}(\nu)\,d\nu \tag{2.4}$$

from which the heat capacity may be calculated if $\rho(\nu)$ is known. An

25

early improvement on the Einstein model was that due to Debye, who assumed that all the modes were like sound waves and $\rho(\nu) \propto \nu^2$ right up to ν_{max}. This theory very successfully explained the T^3 dependence of C_V at low temperatures, and the heat capacity of many solids can be approximated very usefully by the Debye formula with one fitting parameter called the *Debye temperature* $\theta_D = h\nu_{max}/k$. Since ν_{max} is usually about 10^{13} Hz, θ_D is typically a few hundred K. The reason the Debye theory works so well is that at low temperatures the only modes which have $\bar{E}(\nu)$ significantly larger than the zero-point energy are those with low frequencies $h\nu \lesssim kT$, and for these modes $\rho(\nu)$ is indeed proportional to ν^2 as required.

Because its energy is quantized, a mode of oscillation in a crystal can only exchange energy with other modes or some external source of energy such as an electromagnetic wave in units of $h\nu$. A quantum of energy in the lattice vibrations is known as a *phonon*, in analogy to the quantization of an electromagnetic wave into *photons*. Since the lattice vibrations can be described as waves the analogy is quite close, the word phonon obviously originating in the fact that the low frequency waves are sound waves.

In conclusion it is worth repeating that the vibrations we have been considering are harmonic oscillations. This is a very good approximation in practice, but of course the potential energy curves such as in figure 1.3 (*b*) are not exactly parabolic near the potential minimum, and the vibrations are not exactly harmonic; that is they are slightly *anharmonic*. This does not very much affect properties such as the heat capacity, but the anharmonicity is crucially important in causing thermal expansion and limiting thermal conductivity. Thermal expansion arises because the mean position of an oscillating atom is not at the potential minimum if the potential well is anharmonic, and this mean position changes as the amplitude of the oscillations increases with rising temperature. The thermal conductivity of insulators is limited because the carriers of heat, which are the lattice vibrations, interact with one another (see Chapter 4). This interaction would not happen if the vibrations were exactly harmonic.

defects in crystals

3.1. *Types of defects in crystals*

The theory of perfect solids gives a good insight into the general behaviour of crystalline materials, but most real solids depart from perfection in one way or another, and this can have a profound effect on their properties. In this chapter we examine the ways in which crystalline perfection may be lost because of defects in the crystal structure. A defect in a solid is defined as any deviation from the perfectly ordered lattice structure. These defects may be classified into a number of types, and we shall examine them in turn before describing the reasons for their existence and (in Chapter 4) some of their effects.

Point defects

The simplest defect is a *vacancy,* which is a site in the crystal lattice where there is a missing atom or ion. The complementary defect to a vacancy is an *interstitial,* an extra atom or ion inserted at a site where there is not normally an atom or ion of the regular crystal structure. These extra sites are called interstitial sites or interstices. A simple vacancy and interstitial are depicted in figure 3.1. An isolated vacancy

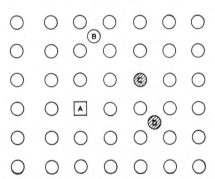

Fig. 3.1. Simple point defects. A: vacancy; B: interstitial; C: substitutional impurity atom; D: interstitial impurity atom.

in a crystal is known as a *Schottky defect,* and the complementary pair of an interstitial and a vacancy is a *Frenkel defect* or *Frenkel pair.* In forming a Schottky defect it is usually assumed that the missing atom is placed on the crystal surface. Under some conditions groups of vacancies or interstitials may gather together to form small aggregates or clusters, for example a pair of vacancies in adjacent sites.

Impurities in solids may also be regarded as defects in the crystal structure. Isolated impurity atoms may occupy *substitutional* or *interstitial* sites as in figure 3.1, but sometimes groups of impurity atoms, or even the precipitation of a separated impurity phase, may occur. Impurities are often referred to as *extrinsic* defects, since they are foreign to the perfectly pure solid. Conversely, vacancies and interstitials are called *intrinsic* defects.

Defects that occupy only a few lattice sites, such as vacancies, interstitials and dissolved impurity atoms are given the generic title of *point defects*. The term is obviously a little vague, but it is useful in practice to distinguish the properties of such defects from those of larger ones.

Line defects

Another important classification is a *line defect*, also called a *dislocation*. Dislocations occur in all crystals and can take on a variety of forms. Figure 3.2 shows a simple *edge dislocation*. It consists of an extra half-plane of atoms inserted into the lattice and terminating at a particular plane, called the *slip plane*. The termination forms a *dislocation line*. It appears as a point on a two-dimensional drawing such as figure 3.2, but is a line perpendicular to the plane of the page in a

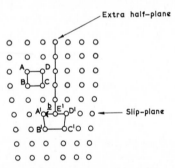

Fig. 3.2. An edge dislocation.

three-dimensional crystal. The lattice planes near a dislocation line bend to accommodate the extra half-plane, so the dislocation introduces a region of strain into the crystal. The region near the dislocation line is known as the *core*. The strain near a dislocation may be expressed in terms of its Burgers vector **b**. This is defined as follows (see figure 3.2). Consider a circuit which in a perfect crystal would be closed, e.g. ABCDA in figure 3.2. If this circuit is taken around the dislocation line, A'B'C'D'E', it fails to close because of the extra half-plane. The closure failure $E'A'$ defines the Burgers vector **b**, whose magnitude

28

represents the amount of strain introduced by the dislocation, and whose direction indicates the directional properties of the strain field (e.g. in figure 3.2 the major strains are obviously sideways). Notice that for an edge dislocation **b** is perpendicular to the dislocation line itself.

Another elementary type of dislocation is the *screw dislocation*. This is a little more difficult to visualize than the edge dislocation, but is illustrated schematically by figure 3.3. There is no extra plane of atoms:

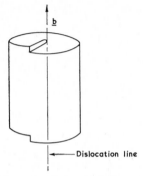

Fig. 3.3.   Schematic illustration of a screw dislocation.

instead a shear distortion is introduced parallel to the dislocation line, so that a circuit around the line takes the form of a helix. Thus the Burgers vector **b** is parallel to the dislocation line. A general dislocation is a mixture of edge and screw components, and hence almost impossible to draw!

Dislocations are extremely important in influencing the mechanical properties of solids. They also have a significant role to play in radiation effects, for a number of reasons. Adding an extra row of atoms to an edge dislocation extends the half-plane further into the crystal: the dislocation is said to have *climbed*. Similarly, removing atoms (for example, by feeding in vacancies to the dislocation) causes climb in the opposite direction. Thus pre-existing dislocations can act as sinks for vacancies and interstitials produced by radiation damage. New dislocations may also be produced by condensation of vacancies or interstitials, but they are usually in the form of *dislocation loops*. These are extra (or missing) planes of atoms which extend through only a limited part of the crystal (typically 10–500 nm). Simple vacancy and interstitial loops are shown in figure 3.4. The boundaries of the extra planes (the dislocation lines) are often approximately circular when viewed from above, hence the term loops. It can be seen from figure 3.4 that vacancy loops involve opposite distortions from interstitial loops. Most solids have preferential planes on which dislocations and dislocation loops lie, resulting from the fact that some dislocation configurations have lower

29

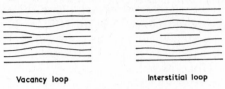

Vacancy loop   Interstitial loop

Fig. 3.4.   Dislocation loops.

energy than others. In diatomic solids such as NaCl and MgO dislocations involve two extra planes, since both types of ion must be present to preserve electrical neutrality.

Grain boundaries

The nature of the boundary between the grains in a polycrystalline material was a question we avoided in the early part of Chapter 2. When there is only a small angle between the crystallographic planes in adjacent grains, the boundary consists essentially of an array of dislocations. The simplest example is the 'pure tilt' boundary shown in figure 3.5, which is an array of edge dislocations. If the dislocations are a distance d apart the angle of this tilt is given by $\theta = b/d$, where b is the magnitude of the Burgers vector of the dislocations. This follows from an inspection of the geometry of figure 3.5. 'Pure twist' boundaries

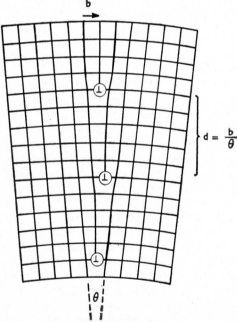

Fig. 3.5.   A small-angle grain boundary, showing how it may be composed of a line of edge dislocations.

30

are also possible with screw dislocations, as are, of course, mixed types.

As the angle of mismatch between adjacent grains widens, the dislocations get closer and closer together. The dislocation model applies up to $\theta \simeq 30°$, but it is clear that for larger angles the separation of the dislocations becomes comparable with the lattice spacing, and the model breaks down. The structure of a large angle grain boundary is thought to be rather like a thin amorphous layer of material between the two crystalline grains. This layer is only a few atomic diameters thick, the atoms in the layer taking up compromise positions to minimize their potential energy. The layer is thin because on either side of the grain boundary the positions of minimum potential energy correspond to the lattice sites of one or other of the crystallites, and there is no tendency to have a gradual, long-range, transition from one orientation of the lattice planes to the other.

Stacking faults

Crystal structures involve the regular arrangement of planes of atoms. In some structures this regular arrangement can be thought of as the stacking of planes of atoms in a definite sequence. Perhaps the simplest example of this is in the face-centred cubic and hexagonal close-packed structures, typical of many metals. The way to pack spheres together so as to minimize the unoccupied volume is in the hexagonal arrangement shown in figure 3.6. If we denote this layer by A, then it can be

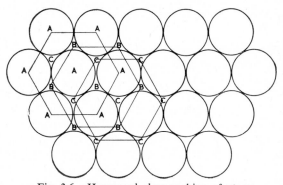

Fig. 3.6. Hexagonal close-packing of atoms.

seen that there are spaces in layer A at positions B and C. Both the lattices B and C are also of the same hexagonal arrangement. To build a crystal with hexagonal close-packed structure, layers of atoms are built up in the sequence ABABAB. . . . To form a face-centred cubic crystal, the sequence is ABCABC. . . . In the latter case the layers A, B and C are (111) planes of the cubic structure (planes perpendicular to the body diagonal of the cube). The energy difference between the two structures

31

is usually quite small (it would be zero if the binding energy depended only on nearest-neighbour bonds, since each atom has 12 nearest neighbours in both structures), and in some circumstances, such as after deformation, the stacking order may be disturbed. Thus a face-centred cubic structure may be disturbed slightly into a sequence such as ABCABABC. . . Such a sequence is said to contain a *stacking fault* at the point at which the sequence is disturbed. A stacking fault is an example of a *planar defect*.

Bulk defects

A further broad classification of defects in solids is into a category which we may call *bulk defects* (sometimes called volume defects). Voids form in some materials as a result of the accumulation of vacancies into large three-dimensional aggregates. Sometimes these voids may be filled with gas, for example, as a result of inert gases being evolved as fission products in reactor materials. The resulting gas-filled void is sometimes known as a gas bubble. The precipitation of impurities into a separated second phase in a solid also gives rise to bulk defects with important consequences, such as the formation of small particles of iron carbide (Fe_3C) in iron, which give steel its mechanical hardness (see Chapter 4). Other examples of bulk defects are pores which tend to form at grain boundaries in polycrystals, and macroscopic features such as cracks and inclusions.

3.2. Origins of defects in crystals
Chemical

Impurities are always present in real crystals. Even a so-called pure crystal of a simple metal, semiconductor or insulator usually contains at least one part per million of impurities, i.e. about 10^{23} impurity atoms per m^3. In some cases great efforts have been put into crystal purification, resulting in impurity levels below this figure. For example, it is possible to obtain silicon crystals with fewer than 10^{18} m^{-3} of certain impurity elements (fewer than one impurity per 10^{10} atoms) because of the emphasis given to the preparation of this material for electronic devices such as the transistor. The presence of impurities in crystals makes an experimenter very wary as to the possible influence of these defects on the results of his experiments.

Impurities are not always a nuisance. In many cases they can be used to manipulate the properties of a material to those desired for some application. We shall see this in more detail in Chapters 4 and 9. As we have remarked before, impurities need not necessarily be present in the lattice as relatively isolated atoms. Often they precipitate into larger particles such as the Fe_3C precipitates in steel mentioned in the previous section. The way that an impurity is incorporated into the lattice depends on its concentration and the temperature. There is usually a limit, called the solubility limit, to the amount of impurity which can be incorporated in solution as isolated atoms; this limit depends on temperature, usually increasing with increasing temperature. At higher concentrations than the solubility limit precipitates of an impurity phase are formed.

Impurities may often give rise to other defects such as vacancies and interstitials. This is particularly marked in ionic and partially ionic crystals, where the atoms (and the impurity) carry a charge. If the impurity has a different valency (and therefore ionic charge) from the host ion that it replaces, then other defects are introduced to compensate the different charge. Consider for example a very common situation, that of calcium impurities in NaCl. The host atoms Na and Cl exist in the solid as singly charged ions Na^+ and Cl^-. On the other hand, calcium is divalent, and its chloride has the formula $CaCl_2$ ($Ca^{2+}Cl^-_2$). Thus for every Ca^{2+} ion entering the NaCl lattice by substituting for Na^+, an extra positive charge is introduced. To compensate this charge (since the crystal as a whole must be neutral) we could introduce an interstitial Cl^- ion for every Ca^{2+} ion, or alternatively leave one Na^+ site empty and so introduce a positive ion (*cation*) vacancy. It turns out that the latter is energetically more favourable (the Cl^- ion is large and there is not much room for an interstitial), so that a crystal of NaCl doped with Ca^{2+} has as many cation vacancies as there are dissolved impurity ions. These vacancies are there simply because of the chemical properties of the impurity. Similar effects are widespread in ionic crystals.

We have not yet said anything about the precise location of the vacancy. Since the Ca^{2+} ion has an extra positive charge and the cation vacancy compensates this, one might expect there to be some attraction between the two. Another way of seeing this is to realize that a cation vacancy represents a missing *positive* ion, and therefore behaves as if it has a *negative* charge with respect to the perfect lattice. It is thus attracted to the site next to the Ca^{2+} ion, with a binding energy which we represent by E_B. If the fractional concentration of Ca^{2+} ions is x and that of the free vacancies (those well-separated from Ca^{2+} ions) is x_v, there must be a fraction $(x-x_v)$ of close impurity–vacancy pairs. The fractional concentration of 'free' Ca^{2+} ions is also x_v. We can regard the association of the vacancy and the Ca^{2+} ion as a chemical reaction:

free vacancy + free Ca^{2+} $\Longleftrightarrow$ *impurity–vacancy pair,*

and the law of mass action then tells us that

$$\frac{x_v^2}{x-x_v} = \tfrac{1}{12}\exp\left(-E_B/kT\right) \tag{3.1}$$

The factor $\tfrac{1}{12}$ comes in because there are 12 equivalent orientations of the impurity–vacancy pair in the NaCl lattice for each cation site occupied by an impurity. Equation (3.1) shows that at low temperatures nearly all the vacancies will be associated with impurities ($x_v \simeq 0$), but that at high temperatures more will become free. E_B is about 0·4 eV for Ca^{2+} in NaCl.

Some compounds form solids in which the law of definite proportions between the constituent elements is not exactly obeyed. For example, nickel oxide (NiO) may contain a slight (up to ~ 2 per cent) excess of oxygen, zinc oxide (ZnO) may contain excess zinc, and uranium dioxide (UO_2) may have excess oxygen. Such compounds are known as *non-stoichiometric* compounds (stoichiometry means having definite proportions between the elements), and usually arise because one of the elements is multivalent, and can form two or more molecular compositions with the other element(s) of the solid. For example, nickel is divalent or trivalent, and can form two oxides, NiO and Ni_2O_3. This causes no trouble in the gas phase, but in a solid it means that the nickel ion can accept varying amounts of oxygen to partner it, whilst still retaining the basic crystal structure appropriate to one of the oxides. Non-stoichiometry is very common amongst compounds of metals of the transition series and the actinides.

Let us examine the case of NiO a little more carefully. The stoichiometry of a sample of crystalline NiO, which has the same lattice structure as NaCl, depends on the conditions under which it was grown. Lightly oxidized pure nickel metal forms stoichiometric NiO, which is green. On the other hand, an oxide layer grown in air at $\sim 1200°C$ has excess oxygen, giving a black material. How is the extra oxygen incorporated? There is not much room for extra oxygen atoms in interstitial positions, so instead the excess of oxygen over nickel takes the form of missing Ni ions, that is Ni vacancies. The story does not end there, however, since each Ni vacancy represents a missing $2+$ ion, which must somehow be charge-compensated. This is done by changing two Ni^{2+} ions into Ni^{3+} for every vacancy. The end result is indeed the formation of some 'Ni_2O_3', but in a way which separates the Ni^{3+} ions throughout the lattice, as shown in figure 3.7. Notice the large

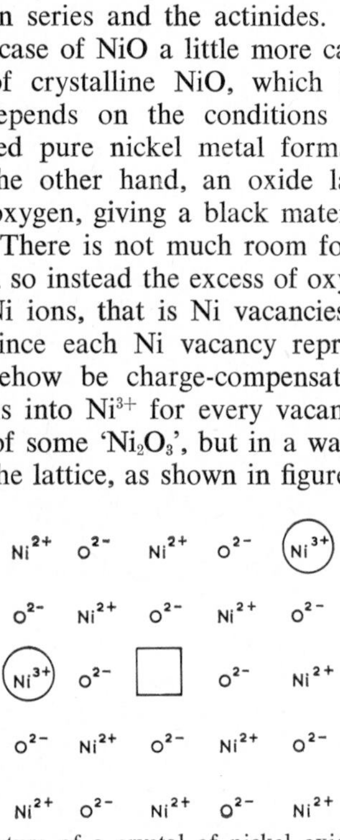

Fig. 3.7. Atomic structure of a crystal of nickel oxide with excess oxygen.

34

change in the detailed structure of the crystal introduced by removing only one nickel ion! Adding Li^+ impurities to NiO also results in the presence of some Ni^{3+} ions for charge compensation.

Similar effects of non-stoichiometry occur in other materials, but the details vary. In UO_2 with excess oxygen the extra O^{2-} ions are incorporated interstitially, since UO_2 (which has the fluorite structure like CaF_2, see figure 2.2) has some large interstitial spaces. In some oxides a precipitate of a higher oxide may occur (for example, Mn_3O_4 in MnO) or the crystal may take up its non-stoichiometry by local changes in crystal structure. Even crystals of compounds where the ions have unique valencies may be made non-stoichiometric by chemical treatment. This is the case for ZnO, where the excess Zn atoms may be incorporated by heating in zinc vapour. Perhaps the most well-known example of this kind of non-stoichiometry is provided by 'additively coloured' alkali halides. If a crystal of KCl (which is normally transparent) is heated at about 600°C in potassium vapour and then cooled quickly to room temperature, it takes on a rather pretty magenta colour. The colour is caused by the presence of F centres (see Chapter 5) (F for the German *Farbe,* meaning colour), which are negative ion vacancies (*anion* vacancies) which have trapped an electron. These trapped electrons absorb green light and hence change the colour of the crystal from transparent to magenta. The F centres are incorporated to take account of the excess potassium introduced during the colouring process. Each extra potassium ion (K^+) in the lattice results in a Cl^- vacancy, and since the potassium is initially adsorbed onto the surface as an *atom,* an extra electron is available which migrates to the vacancy (attracted by the effective positive charge of the anion vacancy). Most alkali halides can be additively coloured to produce F centres, as can some other ionic crystals such as CaF_2, SrF_2 and MgO. The excess Zn in ZnO also changes the crystal colour, from transparent to orange-red, but the details of the way in which the zinc is incorporated have not been definitely established.

If a KCl crystal is cooled slowly instead of quickly from the colouring temperature, it becomes coloured blue rather than magenta. This is because instead of well-separated F centres, a dispersion of colloidal particles of potassium metal is produced. A little thought enables us to see that a precipitation of F centres will give inclusions of potassium metal atoms, which are bulk defects in our terminology. These inclusions scatter light of different wavelengths by different amounts, and hence give the crystal a colour. Similar metallic precipitates can be used to produce coloured glasses.

Thermal

Point defects exist in crystals at finite temperatures for purely thermodynamic reasons, because they represent disorder in the lattice. The second law of thermodynamics gives us the concept of entropy, which represents the amount of disorder in the system. Nature tends to favour disordered systems, since these have the greatest probability of occurrence: there are many ways of making a disordered system, but only a much smaller number of ways of making it ordered. Opposing this chaotic instinct is the unavoidable feature that disorder usually costs energy: an ordered system generally has a lower energy than a disordered one. For example, to make a point defect in a crystal an atom must be pulled out of the structure and placed on the surface or in an interstitial position. This requires enough energy to overcome the binding energy of the atom in its lattice site, so that the disordered crystal has a higher potential energy than the original ordered one. In thermodynamic language, this potential energy difference is the change ΔU in the *internal energy* U of the system. The corresponding change in entropy is usually denoted by ΔS. The actual equilibrium state at temperature T is determined by a compromise between the internal energy U and the entropy S. The quantity TS has the dimensions of energy, and thermodynamics tells us that the state of equilibrium at temperature T is the one with the minimum 'Helmholtz free energy', defined as[†] $F=U-TS$. F is rather analogous to the potential energy of a purely mechanical system, in that it is a minimum at the stable equilibrium state. The equation for F shows that the entropy assumes a more dominant role at high temperatures, but that at low temperatures it is the desire for minimum internal energy which wins. A familiar example is the melting of a solid—an ordered solid with low energy and entropy changes into a disordered liquid with higher energy and entropy at a certain temperature, above which the liquid state is the thermodynamically stable one.

Even though a crystal is a highly ordered state, it can introduce an element of disorder by creating point defects such as Schottky or Frenkel defects. Consider a monatomic crystal containing N atoms and n Schottky defects. The extra internal energy due to the defects is $\Delta U=nE_s$, where E_s is the energy required to form a single Schottky defect. The extra entropy gained may be calculated from the formula introduced by Boltzmann, namely that $S=k \ln W$, where W is the number of different ways of arranging the system. (For a completely ordered state this number is unity, so that $S=0$.) The proportionality constant k is the Boltzmann constant. The problem is thus to calculate W, which we do as follows.

The N atoms and n vacancies can be distributed over $N+n$ lattice sites. The number of different ways W of doing this is ${}^{N+n}C_n={}^{N+n}C_N$, so

---

†Strictly speaking this only applies if the system is kept at constant volume (V). At constant pressure p (which is the normal situation) U should be replaced by the 'enthalpy' $U+pV$.

that

$$W= \frac{(N+n)!}{N! \, n!} \tag{3.2}$$

Thus we may write for the entropy change due to the defects

$$\Delta S = k \ln W \tag{3.3}$$

and

$$\Delta F = nE_s - kT \ln \left[\frac{(N+n)!}{N! \, n!} \right] \tag{3.4}$$

In thermal equilibrium at temperature T, n will be arranged to make the Helmholtz free energy a minimum so that $(\Delta F)/\partial n = 0$. To proceed further we note that in any reasonable solid both N and n will be large numbers ($N \sim 4 \times 10^{28} \, \mathrm{m}^{-3}$), so that we can use Stirling's approximation for $x!$, $\ln x! \simeq x \ln x - x$. Using this formula and differentiating eqn. (3.4), we find that at equilibrium.

$$\frac{\partial(\Delta F)}{\partial n} = E_s - kT \ln \left(\frac{N+n}{n} \right) = 0 \tag{3.5}$$

and

$$\frac{n}{N+n} = \exp \left(-E_s/kT \right) \tag{3.6}$$

For $n \ll N$ this becomes

$$n = N \exp \left(-E_s/kT \right) \tag{3.7}$$

which gives the number of defects present at temperature T. Since E_s is usually about 1 eV and $kT = 0.025$ eV at room temperature ($T = 300$ K), it can be seen that n/N is usually much less than unity. The variation of n/N with temperature for $E_s = 1$ eV is shown in figure 3.8.

When Frenkel defects are considered, we have the problem of arranging n vacancies amongst N lattice sites and n interstitials in N' possible interstitial positions (for most structures $N' \sim N$). Then W becomes (remembering that probabilities multiply each other)

$$W = \frac{N!}{(N-n)! \, n!} \times \frac{N'!}{(N'-n)! \, n!} \tag{3.8}$$

Following through the differentiation of ΔF as before, we find that the number of Frenkel defects is

$$n = (N \, N')^{\frac{1}{2}} \exp \left(-E_F/2kT \right) \tag{3.9}$$

where E_F is the energy required to form a Frenkel pair. Notice that it is $E_F/2$ which occurs in the exponent in this case.

Whether Schottky or Frenkel defects predominate in a particular solid

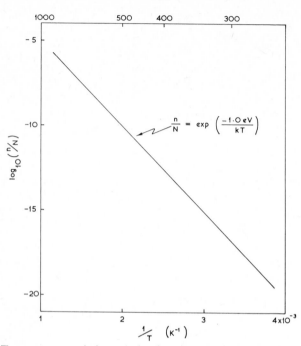

Fig. 3.8. Temperature variation of the function $\exp(-E/kT)$ with $E = 1.0 \text{ eV}$.

depends on the relative magnitudes of E_S and E_F. In most metals, and semiconductors such as Si and Ge, it is found by both theory and experiment (see Chapter 4) that Schottky defects are favoured. This is largely because it costs a lot of energy (several eV) to squeeze an interstitial into most structures.

In ionic crystals Frenkel defects may, in principle, be present independently on the anion and cation sublattices, although in practice one type will predominate by virtue of its lower formation energy. The formation of Schottky defects, however, is constrained by the requirement of electrical neutrality: there must be equal numbers of anion and cation vacancies in pure crystals. A pair of anion and cation vacancies is known as a *Schottky pair*. The number of pairs n_p is given by

$$n_p = N \exp\left(-E_p/2kT\right) \tag{3.10}$$

where E_p is the formation energy of a *pair* and N is the number of ions *of one type*. This follows by writing W as the product

$$W = \left[\frac{(N+n)!}{N!\ n!}\right]^2$$

38

Just like the impurity–vacancy pairs discussed earlier (see eqn. (3.1)), anion and cation vacancies attract one another, and at low temperatures tend to form associated pairs, i.e. vacancies in neighbouring lattice sites. The binding energies in alkali halides are calculated to be about 0·6 eV. As in monatomic crystals, the predominant type of defect found in ionic crystals depends on the various formation energies. In alkali halides the intrinsic defects are Schottky pairs, but in silver chloride (AgCl) and bromide (AgBr) they are Frenkel pairs on the cation sub-lattice, that is Ag^+ vacancies and Ag^+ interstitials. In crystals with the fluorite structure such as CaF_2 and UO_2 the intrinsic defects are anion Frenkel pairs. The cation vacancy is expected to have a large formation energy, since the cations have a higher charge than the anions (U^{4+} in UO_2 as opposed to O^{2-}).

The above arguments have assumed tacitly that the crystal is always capable of attaining thermodynamic equilibrium. As far as vacancies and interstitials are concerned, the crystal is able to do this by a process of *diffusion*. Consider for example the formation of Schottky defects in a monatomic solid, figure 3.9. If an atom A jumps to the

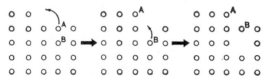

Fig. 3.9. The motion of atoms in a crystal by vacancy diffusion, forming Schottky defects.

surface it leaves a vacancy behind. Atom B may then jump into the vacancy and so on, and in this way vacancies may be distributed throughout the crystal. A similar diffusional process involving jumps of atoms into interstitial sites explains how Frenkel defects may be formed. The process illustrated in figure 3.9 may be visualized as the jumping of a vacancy from one site to the next. The jump frequency p of the vacancy (the number of jumps per second) depends on temperature through an equation of the form

$$p = \nu \exp\left(-E/kT\right) \qquad (3.11)$$

where E is the height of the potential energy barrier an atom has to overcome to move from a neighbouring site into the vacancy (usually about 1 eV), and ν is a characteristic atomic vibration frequency ($\nu \sim 10^{13}$ Hz). Thus for $E = 1$ eV and $T = 1000$ K we have $p \simeq 10^8$ jumps per second, and a jump is occurring quite frequently! The motion of the vacancy as it diffuses through the crystal lattice is a 'random walk', in which each new jump is uncorrelated in direction with the previous one. Consider a vacancy which can move by jumping a distance a in any of three orthogonal directions x, y or z. Let x_i be the distance moved along the x axis in the ith jump (obviously x_i is equal to either $\pm a$ or zero). The distance moved in a total of N jumps will then be

$$x = \sum_{i=1}^{N} x_i \tag{3.12}$$

so that

$$x^2 = \sum_{i=1}^{N} x_i^2 + \sum_{i \neq j} x_i \, x_j \tag{3.13}$$

Since the jumps are uncorrelated, the average value of $x_i \, x_j$ will be zero, so that the mean square distance moved is

$$\overline{x^2} = \sum_{i=1}^{N} \overline{x_i^2} \tag{3.14}$$

We now require the value of $\overline{x_i^2}$. In each jump the vacancy moves a distance $\pm a$, but only one-third of all jumps will be along the x axis. Thus $\overline{x_i^2} = a^2/3$. Furthermore, in time t the vacancy makes pt jumps so that $N = pt$ and the mean square distance moved in the x direction in time t is given by

$$\overline{x^2} \simeq \frac{a^2 \, pt}{3} \tag{3.15}$$

If $p \sim 10^8 \, \text{s}^{-1}$ and $t = 300 \, \text{s}$, $\overline{x^2} = (10^5 a)^2$ and the vacancy may move a linear distance of about 10^5 lattice sites ($\sim 30 \, \mu\text{m}$). This is quite a large distance on the atomic scale, so that diffusion is able to establish thermal equilibrium quite quickly. In forming Schottky defects it may not actually be necessary to put the initial displaced atoms on the surface: internal 'sinks' such as dislocations and grain boundaries always exist, and the diffusion distance may not have to be very large. Equation (3.15) also applies to an interstitial atom jumping at a rate p from one interstitial site to the next.

In some materials, particularly metals with high thermal conductivity, it is possible to 'quench-in' high concentrations of defects by rapid cooling from a high temperature. In this process diffusion is unable to maintain the thermal equilibrium quickly enough, and large numbers of defects are retained in the lattice. The cooled crystal is then in a non-equilibrium condition, rather like the situation of 'frozen-in' disorder in glass. There is a tendency for the solid to get rid of the defects and achieve equilibrium, but the diffusion rate is too slow at the lower temperature for this to happen in reasonable times. For example, at

300 K and with $E=1$ eV, $p\sim10^{-3}$ s^{-1}, and it would take 3×10^{13} s ($\sim10^6$ years) for diffusion to occur over a distance of 30 μm. Contrast this with the situation at 1000 K, where it took only 5 min.

The occurrence of dislocations in crystals is not as easy to understand as is the source of individual point defects. Crystals of Si and Ge grown very carefully contain as few as 10^6–10^7 dislocation lines per m^2, but highly strained materials contain much higher dislocation densities of perhaps 10^{14}–10^{15} m^{-2}. A typical figure for most laboratory grown crystals is 10^9–10^{10} m^{-2}. The number of dislocations present in a crystal depends largely on its history—the conditions during growth, strains introduced during handling, the number of grain boundaries and so on. In principle it ought to be possible to remove dislocations completely by careful annealing. A dislocation line costs about 10 eV of energy for each atomic plane through which it passes, but contributes very little to the entropy of the crystal. This is because the number of ways of putting a line in a crystal is relatively small compared with the corresponding possibilities for point defects described by eqns. (3.2) and (3.8). As a result of the large energy and small entropy that a dislocation introduces, the number of dislocations present in thermal equilibrium should be negligible. Nevertheless it is impossible in practice to remove all dislocations in a solid by annealing, and their presence represents a non-equilibrium situation. This must arise because dislocations are unable to diffuse quickly enough to establish equilibrium, since their motion involves the correlated movement of a large number of atoms, which is very improbable. Another point is that a dislocation can get tangled up with another one, making it difficult for either to move properly out of the crystal. New dislocations may also actually form during the cooling of a sample after annealing or crystal growth because of the mechanical stresses present during cooling and the need to eliminate high concentrations of vacancies. The precipitation of vacancies in a plane may then leave half-planes such as are required to form an edge dislocation (figure 3.2).

Radiation

Irradiation of a solid with energetic particles produces vacancies and interstitials by direct displacement of atoms or ions. The nature of the radiation-damage products forms a wide field of study, since variations in temperature and radiation type have a profound effect on the defects produced. In some solids, notably organic crystals and alkali halides, lattice displacements can be produced by ionizating radiation such as X-rays or even, in some cases, by ultra-violet radiation. The subject of defect production by radiation is discussed more fully in Chapter 7.

CHAPTER 4
defects and the physical properties of solids

4.1. *Dimensional properties*

A perfect crystal has every lattice site occupied by an atom or ion, has no impurities and, if it is a compound, has perfect stoichiometry. The macroscopic volume of the crystal is therefore determined uniquely by the crystal structure and the number of atoms in the solid. Consider for example a simple cubic monatomic crystal with a lattice parameter a. Since each atom occupies the corner of a cube of side a, the total volume of a perfect crystal of N atoms can be made up of N cubes of volume a^3. The total volume is thus $V = Na^3$. If the relative atomic mass (atomic weight) is A_r, the mass of one mole of atoms is $10^{-3} A_r$ kg. The number of atoms in one mole is N_A (the Avogadro constant) which is $6 \cdot 023 \times 10^{23}$ mole^{-1}. So A_r kg contain $10^3 N_A = 6 \cdot 023 \times 10^{26}$ atoms. The density of the solid is then given by

$$\rho = \frac{\text{mass}}{\text{volume}} = \frac{A_r}{10^3 N_A a^3}$$

and is also obviously determined by the crystal structure, in this case by the lattice parameter a. The same general conclusion holds for more complicated crystal structures, except that the geometry is more difficult. A knowledge of the density and the symmetry of the crystal structure can, of course, be used to calculate a value for the lattice parameter.

If the solid is not a perfect single crystal and contains structural defects, then the simple arguments given above no longer hold. The simplest illustration of this is in the case of a polycrystalline material. Very often pores exist at the grain boundaries because of imperfect matching of the shapes of individual grains, and the density of the material will therefore be rather less than the density of the component crystallites. Pores are especially difficult to eliminate when polycrystalline materials are made by pressing or sintering powders together, and the elimination of pores to achieve a high density becomes an important technical problem (see also Section 4.4).

Rather more subtle effects occur in the case of point defects. Consider for example the production of Schottky defects. When an atom is removed from a lattice site and placed on the crystal surface, two things happen. First, the overall volume of the crystal grows because a new atom has been added to the surface. If the volume occupied by one atom in the crystal is V_a (called the atomic volume; $V_a = a^3$ for a simple cubic lattice), then we can write for this contribution to the volume change

$$\Delta V_1 = n V_a \qquad (4.1)$$

where n is the number of Schottky defects.

Secondly, there is generally a small change in position of the atoms near the vacancy, which introduces a strain in the crystal and as a result a further small macroscopic volume change. Let us call this volume change per vacancy V_s. This then contributes a total volume change

$$\Delta V_2 = n V_s \qquad (4.2)$$

to the crystal. The change in linear dimensions of the crystal will then be

$$\frac{\Delta L}{L} \simeq \frac{1}{3} \frac{(\Delta V_1 + \Delta V_2)}{V} = \frac{1}{3} \frac{n}{N} \left(\frac{V_a + V_s}{V_a} \right) \qquad (4.3)$$

because $V = N V_a$. Since the mass of the crystal does not change, the density will change by $\Delta\rho/\rho = -3\Delta L/L$.

Equation (4.3) gives the change in macroscopic dimensions of the crystal. The lattice parameter, however, is not affected by ΔV_1, because this term merely represents the new atoms on the crystal surface. It is affected by the movement of the atoms around the vacancy, because the corresponding strain does introduce an average change in the distance between the atoms in the crystal. This distance may be measured using X-ray diffraction (see Chapter 5), in which the crystal acts as a diffraction grating for X-rays (the wavelength of X-rays is similar to that of crystal lattice parameters, a few tenths of a nanometre). The change in average lattice parameter as measured by X-rays caused by n Schottky defects turns out to be just that expected if the lattice expanded uniformly by ΔV_2,

$$\frac{\Delta a}{a} \simeq \frac{1}{3} \frac{\Delta V_2}{V} = \frac{1}{3} \frac{n}{N} \left(\frac{V_s}{V_a} \right) \qquad (4.4)$$

By combining eqns. (4.3) and (4.4) we therefore have

$$\frac{\Delta L}{L} - \frac{\Delta a}{a} = \frac{1}{3} \left(\frac{n}{N} \right) \qquad (4.5)$$

Thermal expansion, of course, also causes changes in $\Delta L/L$ and $\Delta a/a$. However, these changes are the same for both quantities so that thermal expansion does not affect eqn. (4.5). Combined measurements of the macroscopic length change $\Delta L/L$ and the lattice parameter change $\Delta a/a$ thus give a direct measurement of the Schottky defect concentration (n/N). Both measurements require high precision, because, as we have seen in Chapter 3, n/N is always very small. Figure 4.1 shows results on the thermal generation of defects in aluminium, from which the formation energy of a Schottky defect was deduced (from eqn. 3.7) to be 0·75 eV.

43

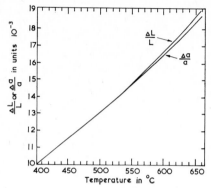

Fig. 4.1. The change in the macroscopic length and lattice parameter of a sample of aluminium as a function of temperature. The difference between the two quantities gives the concentration of Schottky defects (after R. O. Simmons and R. W. Balluffi, *Physical Review*, **117**, 52 (1960)).

Notice that the difference between $\Delta L/L$ and $\Delta a/a$ is small compared with the change in both due to thermal expansion.

How can we be sure that the results of figure 4.1 show that Schottky defects rather than Frenkel defects are formed in aluminium? The distinction between the two is that in the former case the displaced atom occupies a new site on the surface, whereas in the latter it occupies an interstitial position. For a Frenkel defect we therefore have $\Delta V_1 = 0$, because no new lattice sites are created, but we have an additional term

$$\Delta V_3 = nV_i \tag{4.6}$$

because the interstitial causes strains just like those around a vacancy. These strains also contribute to the lattice parameter change so that in the case of Frenkel defects we have

$$\frac{\Delta L}{L} = \frac{\Delta a}{a} \simeq \tfrac{1}{3}\left(\frac{\Delta V_2 + \Delta V_3}{V}\right) = \tfrac{1}{3}\,\frac{n}{N}\left(\frac{V_s + V_i}{V_a}\right) \tag{4.7}$$

The fact that $\Delta L/L > \Delta a/a$ in figure 4.1 therefore *proves* that Schottky rather than Frenkel defects are formed thermally in Al. But in AgBr $\Delta L/L$ and $\Delta a/a$ are found to be equal, which shows that in this material Frenkel defects are the predominant thermally generated ones, as mentioned in Chapter 3.

For a long time there was considerable uncertainty whether irradiation of alkali halide crystals produced Schottky pairs (equal numbers of anion and cation vacancies) or Frenkel defects on the anion sublattice (halogen vacancies and interstitials). The results in figure 4.2 show that irradiated KCl contains Frenkel defects, because $\Delta L/L$ and $\Delta a/a$ are equal at all radiation doses. The halogen vacancies are in the form of F centres (see Chapters 3 and 5), so the irradiated crystals may be compared with

44

those in which the F centres are produced by additive coloration (see Section 3.2). The full line in figure 4.2 shows results for additively coloured KCl, and it can be seen that $\Delta L/L > \Delta a/a$. Since $\Delta a/a$ is positive, it follows that the change in volume per vacancy V_s is positive, so that the F centre occupies more volume than the original Cl^- ion.

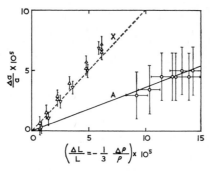

Fig. 4.2.   The lattice parameter change and length change in X-irradiated (curve X) and additively coloured (curve A) KCl (after H. Peisl, R. Balzer and W. Waidelich, *Physical Review Letters*, **17**, 1129 (1966)).

Dimensional changes may also be used to study the influence of impurities on the defect structure of solids. When Ca^{2+} ions are present in KCl the density of the crystal decreases, even though Ca^{2+} is a smaller and slightly heavier ion than K^+. This provides good evidence that the charge compensator for the Ca^{2+} ion is a K^+ vacancy, as discussed in Section 3.2. If it were an interstitial Cl^- ion the density would be expected to increase.

Dimensional changes caused by radiation-induced defects in nuclear reactor components are of crucial technological importance, and will be discussed further in Chapter 8.

4.2.   *Electrical properties*

Metals

The electrical conductivity of a metal or semiconductor is determined by the motion of electrons (or holes) in unfilled energy bands of the crystal. If the electrons were able to move freely throughout the crystal, it would follow that in an applied electric field $\mathscr{E}$ an electron would be given an acceleration

$$\frac{\mathrm{d}\bar{v}}{\mathrm{d}t} = -\frac{e\mathscr{E}}{m_e} \tag{4.8}$$

However, a steady current flow implies a constant drift velocity $\bar{v}$ for electrons, not an acceleration. What happens in practice is that electrons do not move freely through the crystal, but are occasionally deviated or scattered from their paths by some disturbance to the perfect periodic structure of the crystal lattice. In a pure, perfect crystal these disturbances are the lattice vibrations. The result of this scattering is to limit the average time for which acceleration can occur to a 'collision time' τ such that τ is the mean time elapsing between scattering events. The mean drift velocity $\bar{v}$ imparted to the electrons by the field is thus

$$\bar{v} = \frac{\tau}{2} \frac{d\bar{v}}{dt} = -\frac{e\tau\mathscr{E}}{2m_e} \tag{4.9}$$

A rather more rigorous derivation gives a similar result, but without the factor 2 in the denominator, so that from now on we use $\bar{v} = -e\tau\mathscr{E}/m_e$. If there are n conduction electrons per unit volume the current density is $j = n(-e)\bar{v}$, so that $j = ne^2\tau\mathscr{E}/m_e$ and the conductivity $\sigma = j/\mathscr{E}$ is

$$\sigma = \frac{ne^2\tau}{m_e} \tag{4.10}$$

τ is typically 10^{-14} s for a metal at room temperature. For example, in copper $n \simeq 8 \times 10^{28}$ m^{-3}, $e = 1 \cdot 6 \times 10^{-19}$ C, $m_e \simeq 10^{-30}$ kg, $\sigma = 6 \times 10^7$ $(\Omega m)^{-1}$, giving $\tau \simeq 2 \times 10^{-14}$ s.

The conductivity of metals decreases with rising temperature because the increasing amplitude of the lattice vibrations gives increased scattering and hence a smaller value of τ. In fact $\tau \propto T^{-1}$ at high temperatures ($T \gg \theta_D$, the Debye temperature, Section 2.5) and the resistivity σ^{-1} becomes proportional to T. At low temperatures scattering by the lattice vibrations becomes ineffective, and then it is usually found that the resistivity depends severely on the nature of the specimen. Very pure crystals of copper have a resistivity at liquid helium temperatures (4·2 K) of about 10^{-5} that at room temperature, but with standard copper wire a ratio of 10^{-2} is more common. The reason is that the residual scattering at low temperatures is completely controlled by lattice defects such as impurities, quenched-in vacancies and dislocations, which prevent τ becoming very large. A typical figure for the contribution of isolated impurities or of vacancies to the resistivity is about $10^{-6}f$ Ωm, where f is the fractional concentration of defects. A one per cent impurity concentration thus gives a resistivity of about 10^{-8} Ωm, compared with the value $1 \cdot 7 \times 10^{-8}$ Ωm for pure Cu at 300 K, so the influence of defects is obviously crucial. An alloy of one metal with another may be regarded as a metal with a high impurity content $f \sim 0 \cdot 5$, so the higher resistivity of alloys such as brass when compared to pure metals is not unexpected. Figure 4.3 shows the resistivity of copper–gold alloys as a function of composition, showing the maximum resistivity at an intermediate composition.

The increase in electrical resistivity caused by defects is one of the few ways of monitoring point-defect concentrations in irradiated metals. In ideal cases the resistivity per defect can be calibrated by knowing

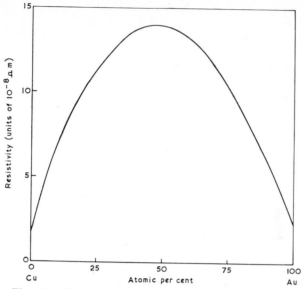

Fig. 4.3. The resistivity of disordered copper–gold alloys.

the absolute defect concentration during other experiments such as illustrated by figure 4.1 in Section 4.1, but usually there is some un-certainty which makes quantitative comparisons with radiation-damage theory difficult. Resistivity measurements are nevertheless widely used and particularly useful in following the annealing of defects as the sample is warmed up.

Semiconductors

Equation (4.10) also describes the conductivity of a semiconductor, except that in general there is also a similar contribution from the valence band holes. An important difference between a semiconductor and a metal, however, is that in the former the number of conduction electrons or holes changes rapidly with temperature. As we saw in Section 2.3, in a pure semiconductor $n \propto \exp(-E_g/2kT)$, where E_g is the band gap between valence and conduction bands. This temperature dependence is much more rapid than that of τ, so the conductivity is controlled largely by n and relatively little by τ. The largest influence of defects on the conductivity of a semiconductor is in modifying the number of conduction electrons or holes, rather than in giving extra scattering. Suppose that a small concentration of phosphorus atoms is introduced into a silicon crystal. The phosphorus impurities substitute for the host silicon atoms, but instead of having four valence electrons they have five. Four of the five electrons participate in the four covalent bonds with the four neighbouring Si atoms and thus form part of the

47

valence band of the crystal. The fifth electron remains bound to the phosphorus atom, but only rather weakly, and occupies an energy level which lies in the gap between valence and conduction bands. Such an energy level is called a localized level because it is associated with a particular impurity site in the crystal. In the case of phosphorus in silicon this localized level is only $\Delta E = 0.045$ eV below the bottom of the conduction band, and electrons may be thermally excited from these levels into the conduction band much more easily than from the valence band, as shown in figure 4.4. Their concentration then varies as

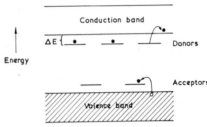

Fig. 4.4.   Donor and acceptor energy levels in a semiconductor.

$\exp(-\Delta E/2kT)$ instead of $\exp(-E_g/2kT)$. Impurities such as phosphorus which provide an *extra* weakly bound electron are called *donors*. Group V elements fall into this class in silicon and germanium. Similarly, impurities from Group III (B, Al, Ga, In) have only three valence electrons and give localized levels just above the valence band which can *accept* an extra electron to complete the four required to make the four covalent bonds. These impurities are therefore called *acceptors*. If an acceptor level takes an electron from the valence band, a hole is left behind, see figure 4.4. This can be viewed as the excitation of a hole from the impurity into the valence band, making the situation complementary to that for donors. The acceptor level of Al in Si is 0.057 eV above the valence band. Point defects produced by irradiation may also behave as donors or acceptors.

The conductivity of a semiconductor may be manipulated over many orders of magnitude by varying the concentration of donor or acceptor impurities. For example, the conductivity of silicon at room temperature may be increased by a factor of 10^3 by having only 10 parts per million of boron acceptors present. A semiconductor containing donors has conductivity by conduction band *electrons*, which are *negative* charge carriers, and is therefore known as *n-type*. Acceptors give conductivity by *positive* valence band *holes* and the semiconductor is then *p-type*. Nearly all the applications of semiconductors in electronic devices like diodes and transistors rely on the manipulation of the conductivity by impurities, particularly by using the properties of p–n junctions. These are regions in a semiconductor where the conductivity changes from

48

n-type to p-type, and they are made by controlling the distribution of impurities in the material (see Section 4.4). A p–n junction acts as a rectifier: it only passes current in one direction. A simple explanation for this may be given by reference to figure 4.5.

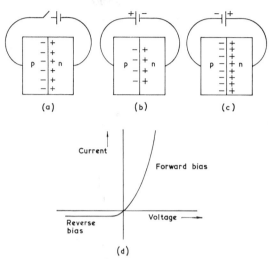

(a) (b) (c)

(d)

Fig. 4.5. The rectifying action of a semiconductor p–n junction. (a) no bias; (b) forward bias; (c) reverse bias; (d) shows the resulting current–voltage characteristic.

When the p–n junction is formed, some intermingling of electrons and holes occurs in the region of the junction: electrons from donors in the n-type region diffuse into the p-type region and recombine with holes there. Similarly holes from acceptors in the p-type region diffuse into the n-type region and recombine with electrons. The recombination of electrons and holes in both regions near the junction gives a *depletion layer* which contains no free charge carriers. It also leaves the acceptor and donor atoms in a charged state, the p side of the junction thus acquiring a negative charge and the n side a positive charge. This establishes an electric field across the junction, opposing further diffusion of charges, and an equilibrium state is reached, see figure 4.5 (a). If a positive voltage is now applied to the p-type side as in figure 4.5 (b), the internal electric field across the junction (which is opposing the flow of charge carriers) is reduced, and a current flows from the p to the n side. This conducting condition is known as *forward bias*. If the voltage is reversed as in figure 4.5 (c) the electric field across the junction is increased, and very little current flows. This is *reverse bias*. The current–voltage characteristics of a p–n junction are as shown in figure 4.5 (d), and demonstrate the rectifying action.

In the metals and semiconductors we have described the charge

49

carriers move freely through the crystal, their drift velocities being limited by the scattering processes determined by the collision time τ in eqn. (4.10). This does not always apply to conduction by electrons and holes, because in some ionic crystals the charge carrier polarizes the surrounding lattice, and by doing so lowers its energy and digs for itself a potential well from which it cannot freely escape. This is sometimes known as *self-trapping*. Self-trapped carriers are very immobile, but can move by jumping over the potential barrier from one lattice site to the next. This process is known as *hopping conduction,* and is thought to occur in some materials such as sulphur crystals. The hopping rate, and hence the conductivity, increases with temperature like $\exp(-E_B/kT)$, where E_B is the height of the energy barrier. The hopping rate is thus like the atomic jumps discussed in Section 3.2 in connection with diffusion.

Insulators

Stoichiometric ionic crystals are good insulators and conduction by electrons or holes is usually negligible. The band gap is so large that even at temperatures close to the melting point very few electrons are excited into the conduction band. For example, it is instructive to compare a typical ionic crystal such as NaCl (band gap $E_g \simeq 5$ eV) at 1000 K with pure Si ($E_g = 1\cdot14$ eV) at 300 K, which is itself quite a good insulator (see Section 2.3).

Using the expression $n \propto \exp(-E_g/2kT)$ (see page 47), and taking the Boltzmann constant k to be $8\cdot6 \times 10^{-5}$ eV K^{-1} (since we are measuring energy in eV), the ratio of the numbers of conduction electrons should be roughly

$$\frac{\exp\,[-5/(2 \times 8\cdot6 \times 10^{-5} \times 1000)]}{\exp\,[-1\cdot14/(2 \times 8\cdot6 \times 10^{-5} \times 300)]} \approx \frac{\exp\,(-29)}{\exp\,(-22)} \simeq 10^{-3}$$

All other factors being equal, the conductivity of the ionic crystal even at 1000 K would be 10^{-3} times that of pure Si at 300 K, that is about 10^{-6} $(\Omega m)^{-1}$. This still represents a very good insulator!

Nevertheless, ionic crystals do conduct electricity at high temperatures. NaCl has a conductivity of $\sigma \simeq 10^{-2}$ $(\Omega m)^{-1}$ at 1000 K as shown in figure 4.6. The conductivity process does not, however, involve the motion of electrons, but instead charged *ions* move through the crystal. Actual transport of matter (the ions) is therefore involved, as may be proved from the fact that electrolysis occurs at the electrodes. This conductivity process is called *ionic* conductivity, and is essentially the same process as occurs in liquid electrolytes. The motion of ions may occur at high temperatures in solids by diffusion, as we noted in Section 3.2. The diffusion of ions can usually only occur, however, if defects are present in the structure, because otherwise there is no possibility of an ion jumping into a neighbouring lattice site. The

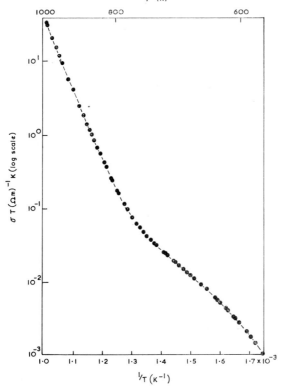

Fig. 4.6. The ionic conductivity of NaCl (after R. Kirk and P. L. Pratt, *Proceedings of the British Ceramic Society*, **9,** 215 (1967)).

important defects depend on the material: in alkali halides the predominant point defects are vacancies, but in silver halides the conductivity is due to the interstitial Ag^+ ions.

The temperature dependence of the ionic conductivity of NaCl is shown in figure 4.6. It is believed that the point defects mostly responsible for ionic conductivity in alkali halides are the cation vacancies. In NaCl, cation vacancies allow the Na^+ ions to jump from site to site and move through the crystal (figure 3.9). There should, of course, also be some anion vacancies present to preserve charge neutrality, but the evidence suggests that the rate at which Cl^- ions jump into Cl^- vacancies is much slower than the jump rate of the Na^+ ions, and the anion contribution to conductivity is small. The conductivity plot in figure 4.6 shows two distinct regions (we shall see later in eqn. (4.17) why $\log(\sigma T)$ is plotted rather than $\log \sigma$), with a sharp change in slope (a 'knee') at about 800 K. The explanation for this is as follows. Below 800 K the number of cation vacancies is determined by the concentration of divalent cation impurities such as Ca^{2+}, and is essentially independent of temperature. The temperature dependence of the conductivity is then controlled only by the jump rate of Na^+ ions into cation vacancies, which is of the form (see eqn. (3.11)).

$$p = \nu \exp\left(-E_M/kT\right) \qquad (4.11)$$

where E_M is called the *activation energy of motion*. The slope of the $\log(\sigma T)$ versus $1/T$ plot in figure 4.6 below the knee then gives $E_M \simeq 0.7$ eV. Above 800 K the number of cation vacancies generated thermally as members of Schottky pairs becomes greater than the number present to charge compensate divalent impurities. One would expect the conductivity to depend on the product of the number of vacancies and their jump-probability, and since the number of vacancies is proportional to $\exp\left(-E_p/2kT\right)$, where E_p is the Schottky pair formation energy (see eqn. (3.10)), the slope of the $\log(\sigma T)$ versus $1/T$ plot above 800 K gives $(\frac{1}{2}E_p + E_M)$. In this way E_p is found to be 2·3 eV. With this information we ought to be able to calculate the concentration of thermally generated vacancies at the knee temperature, 800 K, and check if this is comparable with the concentration of vacancies caused by the divalent cation impurities. From eqn. (3.10) the number of Schottky pairs is

$$n = N \exp\left(-E_p/2kT\right) \simeq 2 \times 10^{28} \exp\left(\frac{-1 \cdot 15}{8 \cdot 6 \times 10^{-5} \times 800}\right) \simeq 2 \times 10^{21} \, m^{-3}$$

This is roughly the expected divalent impurity concentration of high purity NaCl, so the interpretation is self-consistent.

The above arguments show how the temperature dependence of the conductivity is related to point-defect properties, and how electrical experiments give information about the defects. It would be useful if we could also estimate roughly the magnitude of the conductivity, to see if the theory agrees with the experimental values. This would help to convince us that the conductivity of NaCl is indeed due to ionic motion. We can do this, admittedly crudely, as follows.

Consider an ion at A jumping into a neighbouring vacant site at B as indicated in figure 4.7 (a). The jump rate of the ion into the vacancy is given by eqn. (3.11) or (4.11), $p = \nu \exp\left(-E_B/kT\right)$†, where E_B is the height of the potential barrier. If figure 4.7 (a) represents a cross-section through a crystal, there will be $N_o a$ atoms per unit area in a plane passing through A, where N_o is the number of atoms per unit volume (a plane of unit area occupies a volume $1 \times a$ m^3). The average number

---

†In a crystal there will also be a numerical factor (of order unity) to allow for the geometry of the jump. For simplicity we shall ignore this.

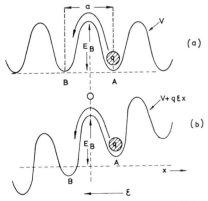

Fig. 4.7. Potential energy for an ion moving in a solid. (a) no electric field; (b) in a field $\mathscr{E}$.

of atoms jumping from plane A to plane B per second is thus

$$\dot{N}_{AB}=N_o afp=N_o af\nu \exp\ (-E_B/kT) \tag{4.12}$$

remembering that a jump can only occur if there is a vacancy at a site in plane B, which we suppose occurs with probability f. Similarly the number jumping from plane B to plane A will be $\dot{N}_{BA}=\dot{N}_{AB}$.

Suppose now that an electric field $\mathscr{E}$ is applied to the crystal as in figure 4.7 (b). If the ions we are considering have charge q, their potential energy will have an additional term $q\mathscr{E}x$, where x is their distance from some arbitrary origin. We may take this to be at the point O without loss of generality. It can now be seen that the energy barrier for jumps from plane A to B is reduced to $E_B - \frac{1}{2}q\mathscr{E}a$, but for jumps from B to A it is increased to $E_B + \frac{1}{2}q\mathscr{E}a$. Thus we now have

$$\dot{N}'_{AB}=N_o af\,\nu \exp\ [-(E_B-\tfrac{1}{2}q\mathscr{E}a)/kT] \tag{4.13}$$

$$\dot{N}'_{BA}=N_o af\,\nu \exp[-(E_B+\tfrac{1}{2}q\mathscr{E}a)/kT] \tag{4.14}$$

We now note that the energy $q\mathscr{E}a$ is much smaller than kT under the conditions of interest. The breakdown field of alkali halides is $\sim 10^7$ V m^{-1} so that $\mathscr{E}$ must be smaller than this, say 10^6 V m^{-1}. With $q\simeq 1{\cdot}6\times 10^{-19}$ C and $a\simeq 6\times 10^{-10}$ m (the lattice parameter of NaCl is $5{\cdot}64\times 10^{-10}$ m) this gives $q\mathscr{E}a\simeq 10^{-22}$ J. Since $k\simeq 1{\cdot}4\times 10^{-23}$ J K^{-1} we have $q\mathscr{E}a \ll kT$ at temperatures of a few hundred K. We can therefore expand part of the exponential terms in eqns. (4.13) and (4.14) to obtain

$$\dot{N}'_{AB}\simeq N_o af\,\nu \exp\ (-E_B/kT)\ [1+\tfrac{1}{2}q\mathscr{E}a/kT] \tag{4.15}$$

$$\dot{N}'_{BA}\simeq N_o af\,\nu \exp\ (-E_B/kT)\ [1-\tfrac{1}{2}q\mathscr{E}a/kT] \tag{4.16}$$

The net current density flowing through the crystal is thus

53

$j = q(\dot{N}'_{AB} - \dot{N}'_{BA})$ so the conductivity $\sigma = j/\mathscr{E}$ is

$$\sigma = \frac{N_0 a^2 f \, q^2 \nu}{kT} \exp\left(-E_B/kT\right) \qquad (4.17)$$

The occurrence of kT in the denominator of eqn. (4.17) explains why (σT) is plotted in figure 4.6 instead of σ.

Let us now put in some numbers typical of NaCl at 1000 K; f is given by $\exp\left(-E_p/2kT\right)$ and $E_B \equiv E_M$ (see eqn. (4.11)) so we can write

$$\sigma = \frac{N_0 a^2 q^2 \nu}{kT} \exp\left[-(\tfrac{1}{2}E_p + E_M)/kT\right] \qquad (4.18)$$

Using $a \simeq 6 \times 10^{-10}$ m, $N \simeq 2 \times 10^{28}$ m^{-3}, $\nu \sim 10^{13}$ s^{-1} and the values previously given for E_p and E_M we obtain

$$\sigma = 2 \times 10^5 \exp\left(\frac{-1 \cdot 85}{8 \cdot 6 \times 10^{-5} \times 1000}\right) \simeq 2 \times 10^{-3} \, (\Omega\text{m})^{-1}$$

Considering that this is only a crude estimate, it is not all that far off the measured value (see figure 4.6).

Ionic conductivity is very widely used to study the properties of point defects in ionic solids, particularly to measure their energies of formation and motion. The results shown in figure 4.6 for very pure NaCl emphasize the extent to which impurities can control the conductivity process. The contribution from less than one part per million of divalent impurities dominates the conductivity below 800 K. Ionic conductivity of real solids at moderate temperatures is nearly always dominated by defects present as charge compensators for impurity ions of different valency, or sometimes by the diffusion of the impurities themselves. In this sense solids are not unlike some familiar liquids: the dangers of having water near electrical components are caused by the conductivity associated with dissolved impurity ions, not the conductivity of pure H_2O itself.

Dielectric loss

In Chapter 3 we mentioned that there is an attractive force between members of a Schottky pair, and also between a divalent impurity and a cation vacancy in an alkali halide, which tend to form vacancy pairs and impurity–vacancy pairs respectively. Although these pairs are electrically neutral and do not contribute to ionic conductivity, they have an effective electric dipole moment and do contribute to the relative permittivity of alkali halides. If an alternating electric field is applied to the crystal the dipoles can follow the changes in field direction only if the pairs can re-orient quickly enough. Since reorientation involves the jumping of atoms into the vacancies, they can only follow alternating fields of frequency ν_{ac} less than the jump rate p. For $\nu_{ac} > p$ the pairs do not contribute to the relative permittivity. When $\nu_{ac} \sim p$ there is a phase lag between the applied field and the polarization of the crystal, and energy is absorbed from the applied field. This is called dielectric loss. Since the jump rate p depends on temperature, measurements of dielectric loss as a function of temperature can be used to measure the energy barrier hindering reorientation of the pairs.

4.3 Thermal properties

Heat capacity

The thermal generation of point defects requires energy, as we saw in Section 3.2. It should therefore give a contribution to the heat capacity over and above that caused by the lattice vibrations, although it will only be substantial at high temperatures where the number of defects is largest. Figure 4.8 shows experimental results for the genera-

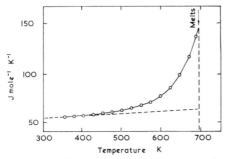

Fig. 4.8.   The heat capacity of AgBr at constant pressure, showing the contribution from the formation of Frenkel defects (after R. W. Christy and A. W. Lawson, *Journal of Chemical Physics*, **19**, 517 (1951)).

tion of Ag Frenkel defects in AgBr, from which their energy of formation can be calculated. The extra internal energy required to form n defects is $\Delta U = nE_F$ so the extra heat capacity at constant volume C_V is

$$C_V = \left(\frac{\partial \Delta U}{\partial T}\right)_V = (NN')^{\frac{1}{2}} \frac{E_F^2}{2kT^2} \exp\left(-E_F/2kT\right) \qquad (4.19)$$

where we have used eqn. (3.9) for n. Experimentally the heat capacity is measured at constant pressure, but the difference between C_p and C_V is very small in a solid and measurements of C_p as in figure 4.8 give a value for E_F of about 1 eV in AgBr.

Thermal conductivity

Electrical conduction is the transport of charge through a material. Thermal conduction is the transport of heat. The heat that is transported is in the form of the thermal energy of the 'particles' which move through the material, and in a gas these particles are obviously

55

the molecules. The kinetic theory of gases gives the following formula for the thermal conductivity K:

$$K = \tfrac{1}{3} C \bar{c} l \qquad (4.20)$$

where C is the heat capacity per unit volume, $\bar{c}$ the average velocity of the particles transporting heat and l their mean free path between collisions. We could introduce a collision time τ as for electrical conductivity (eqn. (4.9) where $\tau = l/\bar{c}$.

In metals the conduction electrons act as the chief carriers of heat, and $\bar{c}$ becomes the velocity of electrons in the conduction band ($\bar{c} \approx 10^6$ m s^{-1} for many metals). Equation (4.20) then shows that it is again the collision processes which determine the thermal conductivity. The situation is generally similar to that for electrical conductivity, and indeed σ and K may be related by comparing eqns. (4.10) and (4.20):

$$\frac{K}{\sigma} = \frac{(\tfrac{1}{3})C\bar{c}l}{ne^2\tau/m_e} = \tfrac{1}{3}\frac{C\bar{c}^2 m_e}{ne^2} \qquad (4.21)$$

The value of $C\bar{c}^2$ for the conduction electrons may be derived using quantum statistical mechanics and is found to be $\pi^2 n k^2 T / m_e$, giving

$$\frac{K}{\sigma} = \frac{\pi^2}{3}\left(\frac{k}{e}\right)^2 T \qquad (4.22)$$

which shows that the ratio of thermal to electrical conductivity is a constant at a given temperature for all metals. This relation is called the Wiedemann-Franz law, and holds quite well in practice. Its main assumption is that the scattering processes which determine l (or τ) are exactly the same for both electrical and thermal conductivity, which is not quite true at all temperatures. Nevertheless, eqn. (4.22) shows that the thermal conductivity of metals is affected by defects in much the same way as the electrical conductivity. At low temperatures the collisions are dominated by defects and σ is constant. In this region K will therefore be proportional to T. At high temperatures scattering by lattice vibrations dominates, and $\sigma \propto 1/T$ so that K becomes constant The general behaviour is shown in figure 4.9. In a pure metal K goes through a maximum at intermediate temperatures; in most metals this occurs at about 20 K. In pure copper the maximum value of K is 5 kW m^{-1} K^{-1}. For impure metals the maximum is suppressed as shown in figure 4.9, because of the increased defect scattering.

In insulators and semiconductors there are very few conduction electrons and the contribution of the electrons to the thermal conductivity is very much smaller than in pure metals. It is then necessary to consider heat transport not by electrons, but by the lattice vibrations. In this case the 'particles' which carry the heat are the phonons discussed in Section 2.4. Heat transport by phonons is the only significant mechanism of heat conduction in insulators and semiconductors. In

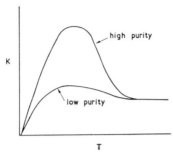

Fig. 4.9. Typical curves for the thermal conductivity K of metals.

metallic alloys the scattering of the conduction electrons is very strong, and the contribution of the electrons to the thermal conductivity is reduced to a size comparable with the contribution of the phonons. However, it is generally difficult to separate the electronic and phonon contributions to K in alloys, so we shall only consider the phonon thermal conductivity of non-metals.

In Section 2.4 we saw that the lattice vibrations could be described as waves moving through the solid, which can only exchange energy in quanta of size $h\nu$, where ν is the frequency of the vibration. These quanta are the phonons. The energy in each vibration is given by eqn. (2.2), so if we equate this to $(\bar{n}+\frac{1}{2})h\nu$, where $\bar{n}$ is now the number of energy quanta (phonons) in the wave ($\frac{1}{2} h\nu$ is the zero-point energy), we find

$$\bar{n}=[\exp (h\nu/kT)-1]^{-1} \tag{4.23}$$

This gives the number of phonons as a function of temperature.

If the lattice vibrations are perfect harmonic oscillations, then a wave can travel through the crystal in an uninterrupted fashion, and in this ideal situation heat (the energy in the wave) could be transported through the solid without resistance. The finite thermal conductivity of phonons arises because the waves are scattered by the crystal boundaries, defects, and, particularly, other waves. In metals they are also scattered by the conduction electrons. The scattering of waves by other waves ('phonon–phonon scattering') comes about because the lattice vibrations are not perfectly simple harmonic. When some anharmonicity is present in the lattice vibrations, the lattice waves are not independent of one another. The anharmonicity causes the presence of one wave to modify the properties of another slightly, and this interaction between the two allows energy to be exchanged between them. This phonon–phonon scattering is the mechanism which limits the phonon mean free path l at high temperatures (as a general rule this means $T > \theta_D$ where θ_D is the Debye temperature). In this region l is inversely proportional to the number of phonons present, just as in a gas the mean free path of

57

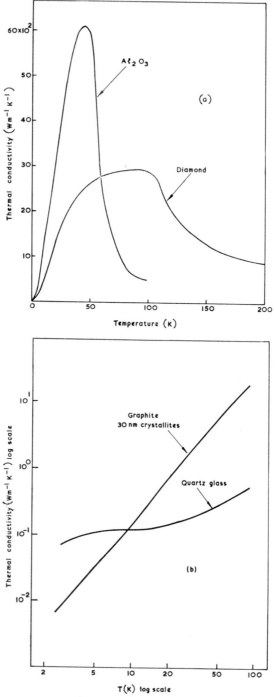

Fig. 4.10. The thermal conductivity of insulators. (a) single crystals; (b) poly-crystalline graphite and fused silica (quartz glass).

a molecule is inversely proportional to the number of molecules per unit volume. From eqn. (4.23), which reduces to $\bar{n} \simeq kT/h\nu$ at high temperatures, we see that $l \propto T^{-1}$ in this case. We may use eqn. (4.20) to describe thermal conductivity by phonons by putting C equal to the vibrational heat capacity (Section 2.4) and $\bar{c}$ equal to the phonon velocity, roughly the velocity of sound. At high temperatures C is constant so that $K \propto T^{-1}$. This is in agreement with experiment. Notice the difference between the high temperature properties of K for metals and insulators: in the former we saw that K becomes constant. The essential difference stems from the heat capacity in eqn. (4.20). The heat capacity of conduction electrons turns out not to be constant at high temperatures, but proportional to T.

In a pure insulator at low temperatures phonon-phonon scattering becomes very weak and the phonons are scattered mainly by the boundaries of the crystal, so that l is independent of temperature and equal to the dimensions of the crystal. Since $C \propto T^3$ at low temperatures $K \propto T^3$. The magnitude of K, however, depends on the size of the crystal! This is a rather strange state of affairs, but it does mean that polycrystalline materials have a much lower thermal conductivity than a single crystal at low temperatures, because the grain size rather than the external dimensions determines l. At high temperatures, once phonon–phonon scattering reduces l to less than the grain size, the difference between the polycrystal and a single crystal disappears. In glasses the mean free path l is of the same order as the inter-atomic spacing. The thermal conductivity of glasses and fine grained polycrystals therefore corresponds to a nearly constant value of l, and K tends to be small and proportional to the heat capacity up to quite high temperatures. Examples of K for single crystals, polycrystals and glasses are shown in figure 4.10.

In an impure or imperfect crystal at low temperatures we might expect, by analogy with the scattering of electrons, that l would be determined by impurity or defect scattering. However, the heat carriers at low temperatures are sound waves of long wavelength, which are generally scattered only very weakly by point defects. Typical wavelengths may be estimated as follows. From eqn. (4.23) a wave of frequency ν only possesses on average a significant number of phonons for $kT \gtrsim h\nu$. The shortest wavelength phonons at temperature T therefore have a wavelength

$$\lambda = \frac{\bar{c}}{\nu} \sim \frac{\bar{c}h}{kT} \qquad (4.24)$$

Putting in the values of h and k, and giving $\bar{c}$ the typical value 5×10^3 m s^{-1} we find $\lambda T \sim 2 \times 10^{-7}$ m. Since the lattice spacing is typically a few $\times 10^{-10}$ m, we see that at low temperatures ($T \lesssim 10$ K) the wavelength is of the order of 100 lattice spacings. Such a long wavelength wave does not 'see' point defects, and passes them unperturbed. Dislocations are a little more effective, but they do not normally affect the thermal conductivity of insulators very much. At higher temperatures phonons of shorter wavelength are present, and point defects have a greater effect. However, by this stage phonon–phonon scattering is becoming important, and the relative effect of the defects is small. There are, however, three cases where defects have a more considerable effect on the thermal conductivity.

A very simple type of defect, which we have not mentioned explicitly before, is an atom of a different isotope of one of the elements of the solid. This defect does not affect the electrons in the solid significantly, but since it has a different mass from the other lattice atoms it does disturb a lattice vibration wave. Many solids contain a mixture of naturally occurring isotopes, and this isotope scattering can reduce the thermal conductivity at intermediate temperatures (that is, temperatures not high enough for phonon–phonon scattering to become dominant). For example, normal germanium is a mixture of 20 per cent ^{70}Ge, 27 per cent ^{72}Ge, 8 per cent ^{73}Ge, 37 per cent ^{74}Ge and 8 per cent ^{76}Ge. The maximum in the thermal conductivity occurs at about 20 K with $K \simeq 1$ kW m^{-1} K^{-1}. Germanium enriched to 96 per cent in the isotope ^{74}Ge has a much higher maximum conductivity of $K \simeq 4$ kW m^{-1} K^{-1}.

Some impurities in insulating crystals have a series of low-lying discrete energy levels which are separated by energies ΔE comparable to the phonon energy $h\nu$ of a long-wavelength lattice vibration. An example is the case of Fe^{2+} impurities on the Zn site in ZnS. The Fe^{2+} ions have a series of five nearly equally spaced levels separated by $\Delta E = 3 \times 10^{-22}$ J (1.9×10^{-3} eV), which result from the perturbation of the lowest atomic energy level of Fe^{2+} by the electrostatic fields in the crystal. The thermal conductivity of this system is shown in figure 4.11, along with that of pure ZnS. The reduction in K at ~ 20 K is caused by a 'resonant scattering' effect when the phonons have the correct energy $h\nu \simeq \Delta E$ to cause transitions amongst the five levels. These phonons are the chief heat carriers at $kT \sim h\nu$, so one would expect the reduction in K to be most apparent at $kT \sim \Delta E$, as is observed ($\Delta E / k = 21.6$ K). At higher temperatures the resonant phonons carry a much smaller fraction of the energy and the effect of the impurities on K is reduced. Similar resonance effects occur for other impurity ions, and also for molecular impurities like OH$^-$ and CN$^-$ in alkali halides, which have low-lying rotational energy levels.

Another case where defect scattering is important is where the defects are comparable in size to the phonon wavelengths at low temperatures. Equation (4.24) shows that this requires defects of dimensions of tens

Equation (4.30) is, in fact, a general relationship between the ionic conductivity and the diffusion coefficient of a particular charged species of concentration N_0. It is called the Einstein relation. The conditions for its validity are often not satisfied in ionic crystals because uncharged species (such as vacancy pairs) contribute to measured diffusion coefficients, but not to the conductivity. We have derived eqns. (4.17) and (4.30) with reference to the specific case of diffusion of lattice atoms with the aid of vacancies (hence the appearance of f in the equations). In Section 3.2 we considered the random walk of a vacancy (figure 3.9), and found that in a time t the mean square distance moved was $\overline{x^2} \simeq \frac{1}{3} a^2 p t$, where p is the jump rate. If we had looked at the diffusion of the *vacancies* rather than the *lattice atoms* in deriving eqns. (4.26)–(4.29), the factor f would have been absorbed in the concentration (remember the definition is flux $= -D \times$ concentration gradient) and we would have found $D_{\text{vacancies}} = a^2 \nu \exp(-E_B/kT) = a^2 p$. Thus we see that, in this case

$$\overline{x^2} \sim Dt \qquad (4.31)$$

Again, it turns out that eqn. (4.31) is a general rule for all kinds of diffusion processes. The numerical factor relating $\overline{x^2}$ to Dt depends on the detailed circumstances under which diffusion is occurring, but is always of order unity.

Diffusion coefficients may be measured experimentally in a number of ways. The diffusion coefficient of an impurity atom is most easily measured by depositing a layer of impurities on the surface of a pure crystal and then leaving the system for a given time at the chosen temperature. The crystal is then cut into thin sections and the number of impurities which have diffused a given distance is measured, for example by chemical analysis. Sometimes a radioactive isotope of the impurity is used and the concentration may then be measured from the amount of radioactivity in a given section. This radioactive tracer method may also be used to measure self-diffusion; for example, the diffusion of Na^+ ions in NaCl may be measured using the radioactive isotope ^{24}Na. Figure (4.12) shows some experimental results for Rb diffusion in K. The plot of log D versus $1/T$ is a straight line as expected from eqn. (4.29).

Very often diffusion plots show several linear regions of different slopes (compare figure 4.6), because various different diffusion mechanisms are taking place. As well as diffusion by mechanisms involving point defects, diffusion processes may take place along grain boundaries or the cores of dislocations. Figure 3.2 shows that there is an expanded region of the lattice near the core of an edge dislocation, down which diffusion may occur more easily than in the rest of the lattice (dislocation pipe diffusion). Similarly grain boundaries involve a region of disorder which assists diffusion. These processes are often important

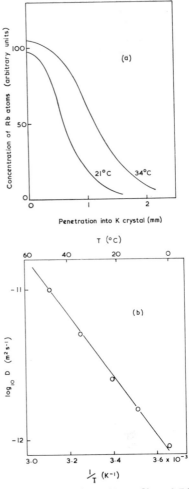

Fig. 4.12. Diffusion of Rb in potassium. (a) profiles of Rb diffused in from the surface; (b) plot of log D versus T^{-1} (after F. A. Smith and L. W. Barr, *Philosophical Magazine*, **20**, 205 (1969)).

at 'low' temperatures (in this context usually room temperature up to a few hundred degrees C) when thermally generated point defects are few and far between.

Diffusion of impurities is used widely in the semiconductor industry, on which so much of modern electronics depends. In Section 4.2 we introduced the p–n junction in a semiconductor (usually Si in modern practical devices), which is the boundary between material doped with different impurities. The way in which a p–n junction is usually made is

64

by taking, say, a slice of n-type silicon and diffusing into it a known amount of a Group III impurity. The diffused region then becomes p-type and the boundary of the region the p–n junction. By controlling the time (remember $\bar{x^2} \sim Dt$) and the temperature (D depends exponentially on $1/T$) the depth of the p–n junction may be chosen to the needs of the device engineer. In modern silicon technology complex microcircuits may be fabricated on a single silicon 'chip' by masking certain regions from the diffusing atoms and carrying out a number of separate diffusion stages (see also Chapter 9).

Diffusion processes are also important in establishing the rate at which chemical reactions occur in solids, or between a solid and a liquid or gas. Two portions of different chemical compounds can only react with each other if diffusion allows the atomic constituents to get at each other. The reaction rate thus increases at high temperatures as more defects are introduced and the diffusion rate of the atoms increases. In a solid–gas or solid–liquid reaction, diffusion in the solid controls the rate at which the reaction proceeds through the solid body. The (relatively!) slow corrosion of iron by rusting is not because of any lack of affinity between iron and oxygen, but because the reaction rate is determined by mobility in the solid state. A better example is the oxidation of aluminium. It is well known that aluminium structures are highly stable in all kinds of weather and do not corrode. Strangely enough, this is not because aluminium does not oxidize easily. In fact quite the opposite: aluminium readily oxidizes to form Al_2O_3, and an exposed clean Al surface quickly becomes covered with a thin oxide layer. However, Al_2O_3 is a strongly bound, stoichiometric, insulating ionic solid, with a high melting point, 2050°C, and the rate of diffusion of oxygen and aluminium through the thin oxide layer is negligible at ambient temperatures. The oxide layer thus protects the aluminium from further corrosion. This is not the whole story, because the mechanical integrity of the film is also very important, as is the question of whether the oxide film covers the whole metal surface: it does not if the oxide occupies a smaller volume than the metal consumed in producing it. The thin film of Al_2O_3 required to stop corrosion of aluminium satisfies both criteria. 'Galvanizing' steel with zinc to produce a layer of zinc oxide has the same effect.

The main problem with the rusting of iron is that the oxide layer required for protection is too thick and breaks off. Stainless steel contains about 8 per cent chromium which helps the formation of a thin Cr_2O_3 film next to the metal which inhibits further corrosion. Iron–aluminium alloys also have good corrosion resistance, but poor mechanical properties. Sometimes, however, a layer of corrosion-resistant alloy can be produced on the outside of a steel component (by diffusion!), which inhibits rusting without significantly reducing the strength. Because diffusion is involved, corrosion processes occur more rapidly at high temperatures, which is a great nuisance in many engineering applications.

In Section 4.1 we mentioned the problem of removing pores in the manufacture of high-density polycrystalline materials. One way in which this may be achieved is by allowing atoms to migrate to the pores by diffusion. Grain boundaries, being regions without regular crystal structure, are able to act as easy sources of atoms for this process, so the pores may be eliminated provided they stay near the grain boundaries. The technical problem is usually that grains tend to grow larger at high temperatures by recrystallization, thus trapping the pores inside large crystallites. This may often be avoided by mixing the material with a small amount of an impurity phase, which precipitates at grain boundaries and inhibits grain growth. Fine-grained, high-density material is then produced by 'hot-pressing': forcing the grains together at a temperature high enough to allow diffusion processes to eliminate the pores.

4.5. *Mechanical properties*

The mechanical properties of solids are discussed in detail in another book of this series, *Strong Materials,* and we shall only consider them briefly here, emphasizing the simpler aspects and the importance of defects. We shall be concerned mainly with the fracture and deformation properties of solids. When a tensile stress is applied to a solid body, we know that a number of things can happen. Following Hooke's law, at low stresses elastic deformation occurs: in other words, there is a small reversible deformation of the material, the strain being proportional to the stress. If the stress is removed, the solid returns to its original size and shape. As the stress is increased, the material either breaks catastrophically at some value of the applied stress, or alternatively the material begins to flow in a plastic-like way. We call the former class of materials 'brittle' and the latter 'ductile'. A ductile material undergoes irreversible deformation. If the material has flowed plastically and the stress is then removed, there is a permanent change in dimensions. The stress at which flow begins is called the yield stress. A ductile material eventually breaks by elongating and 'necking down'; a decrease in cross-sectional area occurs, which in turn causes the stress (force per unit area) to increase. A brittle material, however, breaks before elastic behaviour ceases. If the stress were removed just before fracture occurs, the material would return to its original shape. Our general experience is that metals tend to be ductile, but non-metals brittle. The general situation is shown on a stress–strain diagram in figure 4.13. Obviously these various types of mechanical behaviour should be explicable in terms of the inter-atomic forces. We shall now examine this contention and show that it is only true if defects in the crystal structure are taken into account.

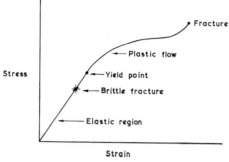

Fig. 4.13. Typical stress–strain curve for a solid.

Brittle fracture

Perhaps the simplest situation ought to be that in which at some critical stress σ_c two halves of a crystal suddenly part company as in figure 4.14. This is essentially brittle fracture. It is usually easier to break a brittle solid body by bending rather than stretching it only because this is a convenient way of applying a high stress. The stress σ_c is usually called the *cleavage stress*: at this stress neighbouring atomic planes in the crystal have just been separated from one another.

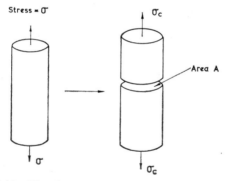

Fig. 4.14. The cleavage of a crystal under tensile stress.

Energy is required to produce cleavage because two new surfaces are made. Atoms on the surface are less strongly bound than those in the bulk, because they have fewer neighbours and hence fewer bonds. Surface atoms therefore have a higher energy than those in the bulk crystal, and as a result it takes energy to create a surface. The extra energy per unit area of a surface is called the surface energy γ, measured in J m^{-2}. It is the solid-state analogue of surface tension in liquids.

We can estimate the surface energy as follows. The creation of two new surfaces of area A as in figure 4.14 requires an energy $2\gamma A$. To do

67

this, a certain number of inter-atomic bonds must be broken. Suppose there are N_o atoms m^{-3} and the lattice parameter is a. There are thus $N_o a$ atoms m^{-2} at the surface. If the bond energy is E_B and i bonds are broken at each surface atom in cleaving the crystal, the energy required is $N_o a i E_B A$. Thus

$$2\gamma A = N_o a i E_B A$$

and

$$\gamma = \frac{N_o a i E_B}{2} \qquad (4.32)$$

For most structures i is a small integer, and $E_B \sim 5$ eV (see table 2.1). With $N_o \sim 4 \times 10^{28}$ m^{-3}, $a \sim 3 \times 10^{-10}$ m we then obtain an order of magnitude estimate for γ of a few $\times 10^{19}$ eV m^{-2} or ~ 1 J m^{-3}. This is a fairly typical value for cases where measurements have been made (γ is 1·2 J m^{-2} for Si, 2·0 J m^{-2} for Fe, 1·2 J m^{-2} for MgO).

What does this tell us about the cleavage stress σ_c? There are a number of ways of estimating σ_c and we follow the simplest. To separate the atomic planes we should have to pull them apart by a distance of order a (the interatomic forces are usually short-range, see Chapter 2). Assuming that the applied stress to do this remains constant at the value σ_c, the work done would be force $\times$ distance, $\sigma_c \times A \times a$. This gives

$$\sigma_c \simeq \frac{2\gamma}{a} \sim 10^{10} \text{ N m}^{-2} \qquad (4.33)$$

An alternative procedure is given in *Strong Materials,* but the deduced values of σ_c are similar to eqn. (4.33).

Measurements of the fracture stress of brittle solids unfortunately give values more like 10^8 N m^{-2}, so there is obviously a large discrepancy between theory and experiment. The explanation of this is that fracture is initiated by cracks in the solid, especially on the surface. A stress concentration may occur at a sharp crack; the actual stress at the tip of the crack is larger than the applied stress by a factor of order $(l/r)^{\frac{1}{2}}$, where l is the length of the crack and r the radius of the crack tip. Cleavage may then occur at a value of the applied stress considerably lower than that given by eqn. (4.33), and once the crack starts to grow ultimate failure is assured, because the stress concentration factor is proportional to $l^{\frac{1}{2}}$. This interpretation of the mechanism of brittle fracture is supported by a number of experimental observations, and applies to amorphous solids such as glass as well as to crystals. The strengths of brittle solids are usually rather unpredictable, which is obviously consistent with the crack mechanism. More convincingly, it is possible to prepare thin fibres of some brittle solids (such as silica, SiO_2) with very perfect crack-free surfaces, and these fibres show fracture strengths approaching the theoretical values.

The surface cracks responsible for the fracture of brittle solids are bulk defects in our classification, and obviously very important ones. They are instrumental in preventing the intrinsic high strength of many brittle solids (such as oxides) from being used in a straightforward way in structural applications. It is worth pointing out, however, that although surface cracks make brittle solids weak in tension, they retain high strength in compression.

Ductility and slip

Plastic flow in ductile solids occurs by a process of slip, which is illustrated schematically in figure 4.15. If the solid is a single crystal the elongation occurs as a result of slip on a number of well-separated planes, the whole section of crystal between adjacent slip planes moving laterally as a unit. In a polycrystal, slip occurs in individual grains, and the only observable effect on a macroscopic scale is the net elongation of the material.

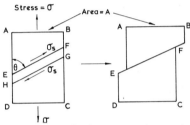

Fig. 4.15. Slip in a stressed crystal.

The solid in figure 4.15 is shown subjected only to a tensile stress σ. How does this cause slip on a plane such as EF? Under the influence of the tensile stress σ there is a force $F = \sigma A$ applied to the solid. Consider a section of material bounded by the planes EF and GH. The component of the force across the surface EF is $\sigma A \cos \theta$ in the direction EF. This 'shearing' force acts in the plane EF and acts across an area $A / \sin \theta$. The shear stress σ_s across the plane is therefore

$$\sigma_s = \frac{shearing\ force}{area} = \sigma \cos \theta \sin \theta \qquad (4.34)$$

The plane GH is subjected to an equal and opposite shear stress, and the net result is a force tending to make the atomic planes in the solid move over one another. If this happens slip occurs on the plane concerned. Slip generally only occurs on certain 'easily sheared' crystallographic planes in a crystal, and for a given plane a critical shear stress is defined as the shear stress across the plane at which slip begins. In figure 4.15 the maximum shear stress occurs across planes at $\theta = 45°$, and has a value $\sigma_s = \sigma/2$. If the solid were a crystal and slip was

69

possible on these planes, a deformation like that shown in figure 4.15 would occur when the tensile stress σ exceeded $2\sigma_{\text{crit}}$. Note that slip is equally possible in tension or compression.

The theoretical critical shear stress of a perfect crystal is estimated in most books on the strength of materials. The argument, originally due to Frenkel, runs as follows. We imagine two neighbouring atomic planes moving over each other as in figure 4.16. Because of the

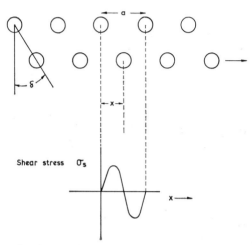

Fig. 4.16. The shearing of ideal crystal planes.

periodicity of the crystal, the force required to move the planes will be periodic, and we assume a sinusoidal variation with the displacement as shown in figure 4.16. The wavelength will be the lattice parameter, a. The force required to shear the planes will be proportional to their area, so it is clearly the shear *stress* which is important and we write

$$\sigma_s = \sigma_{\text{crit}} \sin (2\pi x / a) \qquad (4.35)$$

σ_{crit} represents the stress which has to be applied to cause a permanent displacement, as opposed to an elastic displacement which would return to zero once the stress is removed.

For low stresses we know that elastic behaviour does occur with a shear strain, defined in this case as $e_s = \delta = x/a$ (see figure 4.16), given by

$$\sigma_s = G \, e_s \qquad (4.36)$$

where G is the shear modulus or modulus of rigidity. G is in the range 10^{10}–10^{11} N m^{-2} for most solids. From eqn. (4.35) we see that for small strains x/a we can write $\sin (2\pi x/a) \simeq 2\pi x/a$ and

$$\sigma_s \simeq \sigma_{\text{crit}} (2\pi x / a) \qquad (4.37)$$

70

so that, from eqn. (4.36), we have

$$\sigma_{\text{crit}} \simeq G/2\pi \tag{4.38}$$

The theoretical shear strength of solids is thus 10^9–10^{10} N m^{-2} (note that this is the same order as the cleavage stress). Experiments on ductile single crystals, however, give values some 10^4 lower than this! Again there is a vast discrepancy between theory and experiment, even more so than for brittle fracture.

The high value of the theoretical shear strength may be well illustrated by the following example. The theoretical shear strength of copper is about 10^9 N m^{-2}. The theoretical shearing force required to make a copper wire of 1 mm^2 cross-sectional area flow plastically would thus be 10^3 N which is about the same as the force of gravity acting on a mass of 100 kg. From eqn. (4.34) the tensile force would be about twice this figure. This is equal to the weight of two (heavy!) men. Most of us would doubt the ability of 1 mm^2 copper wire to support such a weight in practice, which emphasizes the difference between theory and experiment.

The observed low critical shear stress of ductile crystals is explained by the presence of dislocations. An edge dislocation (see Section 3.1) can move relatively easily in its slip plane (the plane containing the dislocation line and the Burgers vector), and by doing so can cause slip to occur over a whole section of crystal. This is illustrated by figure 4.17. The arguments leading up to a calculation of the theoretical shear strength required that atomic planes move in unison over one another, which requires a large shear stress. With a dislocation slip is possible without this being necessary. The dislocation only moves one atomic layer at a time, and consequently the shear stress required is small.

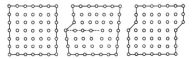

Fig. 4.17. The motion of an edge dislocation in its slip plane, causing slip to occur in a section of crystal (after Taylor).

The stress required to move a dislocation is called the Peierls–Nabarro stress, and in an otherwise perfect ductile crystal is probably of order 10^4 N m^{-2}. This is small enough to allow slip to occur under quite small external stresses, as is observed.

The plasticity of ductile solids may be reduced in a number of ways, which are employed in the manufacture of structural materials. The easy movement of dislocations in the slip plane may be held up by the presence of impurity atoms in the crystal. The dislocation becomes pinned, and requires a high stress to move it. The reason for this is as follows. Isolated impurities in solids tend to segregate near dislocation

71

lines because near a dislocation there are expanded and compressed regions of the lattice which favour the accommodation of a large impurity atom or small impurity atom respectively. Moving the dislocation away from the impurity increases the energy of the crystal, because the impurity site is now less favourable. The increase in energy means that a greater force is required to move the dislocation than in a pure crystal. Defects produced by irradiation can have a similar effect (see Chapter 8). Small particles of a hard impurity phase can also hold up dislocation motion, because it becomes difficult to move the dislocation through or past the precipitate. The most well-known example of precipitate hardening is the increase in strength caused by a fine dispersion of hard Fe_3C precipitates in steel.

Work-hardening is another way of increasing the resistance to plastic flow. The process of slip can itself generate new dislocations, and eventually results in a tangle of dislocations which become more and more difficult to move. This work-hardening is particularly effective in polycrystals because the grains are constrained in their movements by one another, and complex intersecting slip systems can result. Grain boundaries themselves may also act as sources of new dislocations.

All these hardening mechanisms may be used to raise the strength of inherently ductile solids to values which approach the theoretical shear strength. For example, special steels have strengths of up to about 2×10^9 N m^{-2}, compared with a theoretical strength for iron of $6 \cdot 6 \times 10^9$ N m^{-2}. For further details the reader is referred to *Strong Materials*.

Why are some solids brittle and others ductile? The answer lies in the magnitude of the Peierls–Nabarro stress, which is the shear stress required to move a dislocation in the slip plane. If this is large, then plastic flow or slip cannot be achieved easily and brittle fracture aided by cracks is likely to occur before the yield point is reached (see figure 4.13). Conversely, if this stress is low, then slip occurs easily, crack tips are easily blunted by plastic flow, and the experimental cleavage stress increases towards the theoretical value. In a metal the bonding is due to the conduction electrons, and so the inter-atomic forces do not depend strongly on the geometrical arrangement of the atoms. The bonds are 'non-directional'. As a result the changes in atomic positions as a dislocation moves do not cost a lot of energy. On the other hand, the bonds in ionic and covalent crystals depend on the interactions between neighbouring atoms and it is more difficult to move atoms held together by these directional bonds. The Peierls–Nabarro stress thus tends to be high in strongly bonded ionic and covalent crystals such as oxides, diamond and silicon, and small in metals. This explains why the former tend to be brittle and the latter ductile.

In setting out to strengthen a solid there have, of course, to be some compromises. If plastic flow is completely inhibited, then sharp surface cracks cannot be blunted and brittle fracture occurs at a relatively low tensile stress. The strengthening of steels has thus probably reached its maximum practical limit, at which enough plastic behaviour is retained to prevent catastrophic brittle failure. The crack sensitivity of inherently strong, but brittle, solids such as ceramic oxides, glass and graphite can be reduced by 'fibre reinforcement', in which long, thin fibres of the strong material are encased in a ductile high-polymer or metal matrix. The matrix prevents any cracks from running from one fibre to the next, which leads to high strength in a direction parallel to the fibres. (Wood is a very good example of a fibre-composite material). The technique is now being explored widely in materials technology, fibreglass probably being the best known man-made example.

At high temperatures, where diffusion can occur at an appreciable rate (usually $T \gtrsim 0.5 \times$ melting temperature in K), all solids undergo a slow time-dependent deformation under stress known as *creep*. The mechanisms of creep are generally complex. They may involve the motion of dislocations because pinning becomes ineffective at high temperatures on account of the dissolution of precipitates or the diffusion of impurities. The annealing-out of dislocation entanglements which gave work-hardening, the movement of grain boundaries (recrystallization or sliding) and the transport of lattice atoms by diffusion are other possibilities. Creep-resistant materials are essential for components working at high temperatures such as gas-turbine blades. Materials for these applications must have small diffusion rates and special alloys have been developed for this purpose. We shall discuss briefly the importance of creep in nuclear reactor components in Chapter 8.

CHAPTER 5

interaction of solids with low energy radiation

5.1. *Interaction with optical and infrared photons*

If we look at solids in the world around us it is apparent that there are four groups into which we can usefully classify them as far as optical properties are concerned. First there are solids like glass and Perspex. They are completely colourless and very transparent to visible light; they do reflect some light from their surfaces but they are not particularly good reflectors. Secondly, there are solids which are also transparent and are like the first group in most respects except that they are coloured. Coloured glasses and plastics are the most obvious examples, but large numbers of relatively pure chemical compounds, such as copper sulphate crystals, also have these properties. The third group includes opaque materials like brick, wood and matt paint. They do not reflect light regularly as do mirrors or window-panes, but they do reflect diffusely. Finally, there are metals, which are completely opaque, even in relatively thin layers, and are also bright and very good reflectors. In many cases they are not coloured, having a 'silvery' look, but often they are: for example copper and gold.

In some respects this division of solids into four groups is rather artificial. After all, if glass is ground very finely into a powder and then compacted it looks just like a white brick; certainly it does not transmit light or reflect it as a window-pane does. Furthermore, if we had eyes which were sensitive to ultraviolet or infrared radiation, instead of just to visible light, then Perspex would look much more intensely coloured than copper sulphate. In these respects our first three groups are one, and only metals truly differ, but by keeping all four we can more smoothly introduce the important physical processes such as the absorption and scattering of light.

Clear dielectric solids

A block of glass or Perspex with a polished surface does very little to visible light except to reflect some of it from the surface and to bend (refract) rays of light which do not strike the plane surface normally. We say it is a clear *dielectric* medium, through which the light can pass unaffected except for its reduced velocity. A light wave has electric and magnetic fields oscillating normal to the direction of propagation. The frequency (ν) of these oscillations and the wavelength (λ) are related to the wave velocity (v) by the equation

$$\nu\lambda = v \tag{5.1}$$

In a vacuum the velocity of the wave is c ($=3 \times 10^8$ m s^{-1}), and in a transparent solid it is slower by a factor n, which is called the *refractive index* of the solid.

The physical reason for this slowing down is as follows. When a light wave passes through a solid, its electric field produces an electrical polarization, that is electrons are forced to oscillate relative to their parent nuclei by the oscillating electric field of the light; if the frequency of the electromagnetic wave is low enough positive ions can also move relative to negative ions, and even whole molecules can reorientate in the field. Because of this forced oscillation of electrons and atoms inertia is added to the response of the medium to the electromagnetic wave and the wave propagation velocity is reduced. A mechanical analogy is a vibrating string which is being excited at a constant frequency; if masses are added to the string the wavelength and wave velocity are reduced. In magnetic solids there is an additional effect because of the extra magnetization caused by the oscillating magnetic field of the light wave, which adds even more inertia to the system.

Maxwell, in the nineteenth century, was the first to develop an adequate mathematical formulation of these ideas. He showed that the space and time variations of the electric and the magnetic field of the radiation are related by the *wave equation* which we write simply here as

$$\frac{\partial^2 \mathscr{E}}{\partial x^2} = \epsilon\mu \frac{\partial^2 \mathscr{E}}{\partial t^2} \; ; \; \frac{\partial^2 H}{\partial x^2} = \epsilon\mu \frac{\partial^2 H}{\partial t^2} . \tag{5.2}$$

In this equation the response of the medium to the electric field is given in terms of the permittivity, ϵ, and to the magnetic field in terms of the permeability μ. The resulting wave velocity is $(\epsilon\mu)^{-1/2}$.

For most clear dielectric solids like glass and Perspex μ is not very different from μ_0. Since ϵ is normally given in terms of the relative permittivity ϵ_r ($=\epsilon/\epsilon_0$) the wave velocity in these solids is $v = (\epsilon_0\epsilon_r\mu_0)^{-\frac{1}{2}} = c/\sqrt{\epsilon_r}$. This explains the commonly used approximation for refractive index, that it is equal to the square root of the relative permittivity. In fact the approximation is actually valid only for non-magnetic materials where $\mu \simeq \mu_0$.

At the very highest frequencies, corresponding to X-rays or γ-rays, even atomic electrons have too much mass to move in response to the oscillating electric field and $\epsilon_r \sim 1$. At optical frequencies, on the other hand, electrons can move quickly enough for forced oscillation to occur and ϵ_r is usually in the region of 2–3 (for NaCl $\epsilon_r = 2.25$). At even lower frequencies ions can move and ϵ_r becomes larger still, eventually reaching the zero frequency value which is written ϵ_{ro} (for NaCl

75

$\epsilon_{ro}=5.62$). In molecular crystals such as solid CH_3Cl where the molecules have an electric dipole moment the reorientation of the whole molecule by the field can contribute to ϵ_{ro}. As a result of all these effects the relative permittivity of a solid usually varies with frequency as shown in figure 5.1. Later we will deal more fully with the 'absorption' which is related to the odd behaviour at the transitions between regions, but the physical basis for the anomalous behaviour is not hard to understand. For example, in figure 5.1 we have assumed

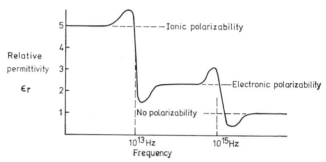

Fig. 5.1. Relative permittivity of an ionic solid as a function of frequency.

that at frequencies well below 10^{13} Hz the ions are able to move in response to the oscillating electric field and so contribute to the effective inertia of the medium. As the frequency of the incident wave is increased it eventually approaches the natural vibrational frequency of the ions in the crystal. For any system undergoing forced oscillations, resonance occurs when the frequency of the forcing disturbance is the same as the natural frequency of the system. So, around this frequency the ions in the oscillating electric field vibrate with greatly enhanced amplitudes. The contribution to the inertia can then be much larger. At higher frequencies still the ions are unable to follow the oscillations of the electric field, the ionic contribution disappears altogether and the refractive index falls to that given by electronic motion alone.

Optical absorption in pure insulating solids

If some of the light passing through a solid is to be absorbed it must be possible to transfer some energy from the wave to the solid. In the quantum theory a light wave can give up energy only in definite amounts, indeed a beam of light is often imagined as a stream of particulate *photons* rather than a wave. The energy of a photon is given by

$$E = h\nu = hc/\lambda \qquad (5.3)$$

and if E does not correspond to the spacing of some energy levels of electrons in the solid, then it is not possible to transfer energy from the light to the electrons. In Chapter 2 we saw how insulators have completely filled valence bands and empty conduction bands separated by a band gap E_g. For a photon to be absorbed we must therefore have $h\nu > E_g$ so that an electron can be excited across the gap from valence to conduction band. If $h\nu < E_g$ the photons cannot be absorbed and the light will pass through the solid as through a clear dielectric, suffering only a reduction in velocity as we discussed earlier. For glass $E_g \sim 4$ eV so that $E_g/h \sim 10^{15}$ Hz, which means that even violet light ($\nu = 6 \times 10^{14}$ Hz) cannot be absorbed.

If the wavelength is small enough, so that ν is large enough, then photons can be absorbed and give up their energy to the solid. The probability that a given photon will be absorbed in passing through a thin layer of thickness dx is usually written $\alpha\,dx$, where α is called the *absorption coefficient* of the material. In passing through a layer of finite thickness x the intensity of an incident beam falls from I_0 to I where

$$I = I_0 \exp(-\alpha x) \tag{5.4}$$

Corrections for reflection by the surfaces may have to be made, because then some of the light may traverse the material several times. The formula in this case is given in figure 5.2. At wavelengths where electrons can be excited from the valence band to the conduction band, solids such as single crystals of common salt or plastics have absorption co-efficients up to 10^7 to 10^8 m^{-1}, which means that layers only 10–100 nm thick will absorb a large fraction of the incident light. In some crystals exciting electrons from one band to another is not quite so easy, the transition is said to be partly forbidden and the absorption coefficient is therefore much smaller.

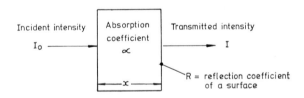

$$I = I_0(1-R)^2 \exp(-\alpha x)/[1 - R^2 \exp(-2\alpha x)]$$
$$\sim I_0 \exp(-\alpha x) \text{ if } R \text{ is small.}$$

Fig. 5.2. The effect of reflection and absorption on the transmission of light through a crystal.

If the energy required to excite an electron corresponds to a visible photon, then the crystal will look coloured. Organic dyes have absorption bands in the visible region and in thin layers they are brightly coloured. If, as in figure 5.3, the dye absorbs the violet, blue, green and

77

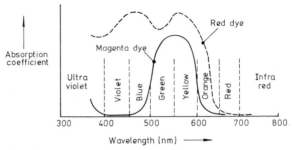

Fig. 5.3. Absorption coefficient of red and magenta dyes as a function of wavelength.

yellow parts of the spectrum, only the red components of incident white light remain, and a fabric in which small amounts of dye are incorporated will look red. Similarly, if the dye absorbs in the green and yellow region, then violet-blue and red-orange light will not be absorbed as light is scattered by the dye-impregnated fibres and the fabric will have a magenta colour. Although their absorption of light is selective, as shown in figure 5.3, dyes usually also have significant absorption coefficients throughout the whole of the visible region of the spectrum and as a result thick layers of dye crystals or concentrated solutions look black. Crystals like copper sulphate also have absorption bands in the visible but the transitions are partly forbidden, so that the absorption coefficients are much smaller and even thick crystals have a coloured but transparent appearance.

As one might expect, the transmission of light through solids is obviously and considerably changed by absorption processes. The reflection of light from the surface of a solid is also modified by the presence of absorption bands. The reflection coefficient at the surface between two dielectrics for light incident normal to the surface is

$$R = (n_2 - n_1)^2 / (n_2 + n_1)^2 \qquad (5.5)$$

At a plastic–air interface this becomes $(n-1)^2/(n+1)^2$ and for most plastics $n \sim 1.3$ so that about 2 per cent of the incident light is reflected. However, for light passing from air to a solid with refractive index n and an absorption coefficient α the fraction reflected becomes

$$R = \frac{(n-1)^2 + (\alpha\lambda/4\pi)^2}{(n+1)^2 + (\alpha\lambda/4\pi)^2} \qquad (5.6)$$

If we again use $n = 1.3$ but with $\alpha = 5 \times 10^7 \, \text{m}^{-1}$ and $\lambda = 500$ nm the reflection coefficient becomes about 70 per cent, compared with 2 per cent for $\alpha = 0$. There are two obvious results of this effect, the first being that solids which absorb very strongly in the visible region also reflect well and therefore often have a shiny black appearance. The

78

second point is that reflection coefficients are high where absorption coefficients are also high, so that dye crystals look one colour in reflection and the complementary colour in transmission. A fairly commonplace example of this is red ink dried on a pen-nib, which usually looks green. The red dye has an absorption coefficient like that sketched in figure 5.3. Viewed in transmission it will look red, with only red light being transmitted. In contrast, if white light is reflected from its surface the reflection coefficient will be much higher for blue, green and yellow light than for red, and it will look green.

Photons of infrared radiation rarely have enough energy to excite electrons in pure insulators. They can, however, excite lattice vibrations if the wavelength is appropriate. In common salt and other alkali halides the positive and negative ions oscillate naturally in the lattice at frequencies above 10^{13} Hz, corresponding to wavelengths of 20–100 μm. As a result such infrared radiation is strongly absorbed by these crystals. The photons are absorbed and vibrational quanta (phonons) are created in the process. The absorption coefficients can reach a few times 10^6 m^{-1}, which is rather less than for electronic transitions because the masses of the excited particles are so much greater. The frequency at which absorption occurs in diatomic ionic solids is also related to their masses, by the expression

$$v \simeq \frac{1}{2\pi} \left[2g \left(\frac{1}{M_1} + \frac{1}{M_2} \right) \right]^{\frac{1}{2}} \tag{5.7}$$

in which M_1 and M_2 are the masses of the two ions and g is the force constant for the bond. It is clear that v can be made small, giving good transmission through most of the infrared region, only by making both M_1 and M_2 large. Thus LiF transmits infrared radiation only at wavelengths below 5–6 μm, whereas KCl transmits to about 20 μm and KRS-5, a commercial infrared window material which is a mixed thallium bromide-iodide crystal, transmits usefully out to about 50 μm.

Absorption by defects

Pure single crystals of alkali halides are all clear transparent solids without any colour at all. However, many of them become very brightly coloured when irradiated with X-rays or some other kind of high energy radiation, indicating that some defect which absorbs visible radiation has been produced. Figure 5.4 shows three samples of sodium chloride (common salt). The left-hand sample is a relatively pure single crystal and is a clear dielectric solid. The centre sample has been irradiated with X-rays; in a colour print it would be brown-yellow but in this print it is dark grey. Crystals coloured in this way typically have peak absorption coefficients of about 5×10^3 m^{-1}, which is, of course,

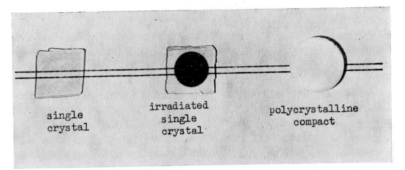

single
crystal

irradiated
single
crystal

polycrystalline
compact

Fig. 5.4. The transmission of light through sodium chloride. (*a*) a single crystal; (*b*) a single crystal which has been irradiated; (*c*) a polycrystalline compact.

many orders of magnitude less than for coloured dyes. Coloured gemstones are similar to irradiated alkali halides in that the optical absorption is also due to relatively small concentrations of defects, in this case impurity atoms, in an otherwise colourless, transparent host.

In terms of the band picture this effect is very easy to describe. The defects introduce electronic levels in the band gap as shown in figure 5.5 just as did the donors and acceptors in semiconductors discussed in

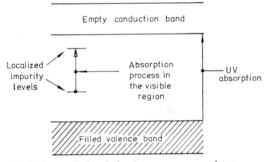

Fig. 5.5. Localized electronic levels in the energy gap of a transparent insulating crystal, causing visible absorption processes.

Chapter 4. Transitions between these levels cause the visible absorption. The absorption coefficient of a solid which is caused by defects is proportional to the concentration of the defects, so that lightly irradiated alkali halides are scarcely coloured at all and heavily irradiated ones are nearly black. F centres are the defects which primarily cause the colour in irradiated alkali halides, and they consist of negative ion vacancies which have trapped electrons (figure 5.6). That part of the crystal from which a negative ion is removed behaves as if it is positively charged, and can bind an electron to it just as a proton can.

In fact, one crude but useful analogy of an F centre is that of a hydrogen atom embedded in a dielectric.

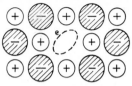

Fig. 5.6. Atomic structure of the F centre in alkali halides.

It is probably already apparent that measurements of optical absorption as a function of wavelength can provide very useful information about defects in insulators. Sometimes the position and shape of the absorption band can help in identifying the microscopic structure of the defect; always the absorption coefficient can be used as a convenient measure of the concentration of the defect and this is often very valuable.

Just as pure crystals can absorb electromagnetic radiation by the excitation of atomic vibrations as well as by electron excitation, so can defects. One defect where the absorption of infrared radiation is very marked is the so-called U centre. This has the structure of an F centre to which a hydrogen atom has been added; it is an H^- ion in a negative ion vacancy. Because the hydrogen atom is so much lighter than the alkali or halogen atoms its natural vibration frequency is much higher, and whereas KCl itself absorbs infra-red radiation most strongly at a wavelength of about 50 μm, U centres in KCl absorb in a strong sharp line near 20 μm (figure 5.7). Hydrogen-containing centres give the best examples of the way in which the mass of the vibrating ion is related to its absorption frequency. Light hydrogen atoms can be replaced by deuterium, which have double the mass but otherwise identical chemical properties, and then the frequency of absorption is reduced by $\sqrt 2$ as shown in figure 5.7.

Dielectric scattering

Most pure insulating solids, such as alkali halides, aluminium oxide and silica, are perfectly clear and transparent to visible light provided they are in suitable form. Most inorganic insulators in the everyday world around us, for example, porcelain and brick, are not clear and transparent. Why is this?

The primary reason is that everyday ceramics are not single crystals, nor even glasses, but polycrystalline compacts. Polycrystallinity was briefly mentioned in Chapter 2 (Section 2.1); instead of forming a single crystal or at least a continuous phase the solid has been formed as many tiny crystals packed together in a random array. These crystal grains are separated by interfaces called grain boundaries, where the

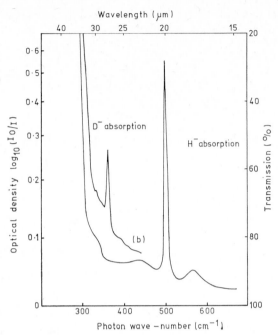

Fig. 5.7. Infrared absorption bands of U centres (substitutional H⁻ centres) and D⁻ centres in KCl (after A. E. Hughes and D. Pooley, *Solid State Communications*, **10**, 53 (1972)).

atoms in the crystal have a slightly different structure and density from the bulk. In most practical cases the micro-crystals do not join together perfectly and pores form at the grain boundaries (see fig. 2.3). In either case refractive index changes occur at the boundaries and light is scattered there by virtue of eqn. (5.5). As little as 1 per cent of the incident light may be scattered but, if the grain boundaries are sufficiently numerous, the solid will still look opaque. A practical example of this scattering effect on a large scale is when a motor-car windscreen shatters. Thousands of small sections of glass are produced and, although each is as clear and transparent as ever it was, the scattering at the boundaries prevents clear vision through the windshield. The right-hand sample of sodium chloride in figure 5.4 also illustrates the effect.

Scattering at interfaces is the main reason why white pottery is not transparent. However, most commercial ceramic materials either are accidently not very pure or else they are deliberately loaded with impurities to give them a colour. The colour-forming process is just as in crystalline or glassy solids, that is by straightforward optical absorption, but here it is added to the scattering discussed above. The

82

combination of scattering and absorption gives most of the common materials around us their characteristic appearance; this is why fabrics and wood look as they do. One of the most interesting examples of combined absorption and scattering effects is in paints. The basis of gloss paints is usually some organic resin, which alone would be water-proof but would not be very attractive and would both discolour in sunlight and fail to protect the underlying wood from the sun's radiation. In white gloss paint small particles of TiO_2 are added. Although TiO_2 is also colourless and transparent, its refractive index is very different from that of the resin base (2·7 as against 1·3) so that a lot of scattering at the interfaces between TiO_2 and resin occurs, giving the paint an attractive white brightness and ensuring that the sunlight does not reach the wood or even the deeper layers of the organic base. In coloured paints impurities or dyes are added, but the scattering remains, and most of the light incident on the surface is reflected.

White paints are particularly good reflectors of incident light, but the light is obviously not reflected as in a mirror. If we shine a narrow, collimated beam of light from a laser onto a flat paint surface we find that the reflected light leaves the surface at a variety of angles as shown in figure 5.8. This is called *diffuse reflection* and is obviously very different in character from the *specular reflection* which occurs at the surface of a metal. The diffuseness of diffuse reflection is not really surprising, after all we are relying on reflection from a very large number of randomly oriented particles. It is perhaps not so easy to see why the surfaces of metals are such good reflectors.

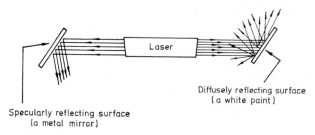

Fig. 5.8. Specular and diffuse reflection of laser light.

Metals

A sheet of polished glass reflects only about 2–3 per cent of the light incident normally on it. In contrast a sheet of polished silver will reflect 94 per cent of the same light beam. This is a striking difference, but it should not really surprise us since we have already noted that strongly absorbing compounds like dyes are also good reflectors.

According to the electron band description of solids metals differ from insulators and semiconductors in having a conduction band partially filled with electrons. In metals, therefore, there are unoccupied electron states just higher in energy than the uppermost filled ones, and because of this the electrons can re-arrange themselves in an applied electric field to give a current flow. However, we saw in Chapter 4 that the moving electrons are scattered from time to time by lattice vibrations and defects, and in the scattering process lose some of their kinetic energy, which becomes lattice vibrational energy. In this way they take energy from the electric field and convert it into heat. In the same way light is absorbed by metals; the electric field of the electromagnetic wave gives up energy to the electron and this is transferred to the lattice vibrations.

Insulators obviously do not absorb radiation in this way, so it is perhaps not surprising that the absorption coefficient of a metal should increase with increasing conductivity. Maxwell's equations show that the absorption coefficient is given by

$$\alpha = 2(\pi\mu\sigma\nu)^{\frac{1}{2}} \qquad (5.8)$$

where μ is the permeability and σ the conductivity of the metal. For copper at room temperature $\sigma = 6 \times 10^7$ $(\Omega m)^{-1}$, so that for 1 μm radiation ($\nu = 3 \times 10^{14}$ Hz) the absorption coefficient is about 3×10^8 m^{-1}. This means that the light is absorbed in a small fraction of its free space wavelength; the absorption coefficient is extremely high. If we put the value of absorption coefficient calculated from eqn. (5.8) into eqn. (5.6) for reflectivity, and also assume that $n \sim 1 \cdot 5$ for copper, we have a reflectivity of about 99 per cent. This is rather more than is actually obtained, but the reflectivity of many metals in the infra-red does reach 95–99 per cent.

In order to be a good reflector the metal film must be much thicker than the characteristic distance $2/\alpha$, which is called the *skin depth* of the metal for the frequency in question. The skin depth increases with decreasing frequency, as shown in table 5.1. It is apparent from this

Table 5.1. The skin depth in copper at room temperature at a variety of frequencies

Radiation frequency	Free space wavelength	Region	Skin depth
3×10^{14} Hz	1 μm	Infrared	4 nm
300 GHz	1 mm	Short microwaves	120 nm
300 MHz	1 m	UHF broadcasting	4 μm
300 kHz	1 km	Longwave radio	120 μm
300 Hz	1000 km	Audio	4 mm

table that quite thick metal shields are needed to reflect electromagnetic waves at low frequencies; several cm of copper would be needed to shield electronic equipment from 300 Hz radiation. Since the absorption coefficient increases with magnetic permeability μ shielding can often be improved by using a magnetic alloy.

The absorption coefficient given by eqn. (5.8) is quite close to the experimental one for infrared and lower frequencies, where the energy which can be given to the electrons by the electromagnetic wave is small compared with the width of the conduction band. At optical frequencies and higher this is no longer true and the absorption coefficient deviates from eqn. 5.8. In some metals, such as gold and copper, there are higher values of the absorption coefficient in some parts of the visible region than in others. As a result of eqn. (5.6) the reflectivity is higher for some wavelengths than others and the metal appears coloured. This is illustrated in figure 5.9.

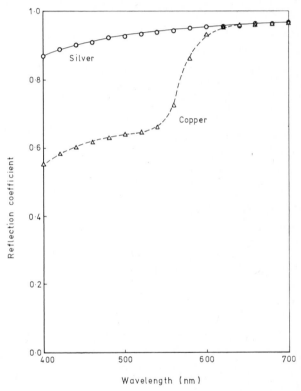

Fig. 5.9. Reflectivity of silver and copper as a function of wavelength (courtesy of G. P. Pells).

The optical and infrared properties of metals are determined by the very large number of conduction band electrons, and it is therefore very difficult to detect the small optical changes that are brought about by introducing defects. Because of this, optical studies of defects in metals

85

are usually not very fruitful, whereas they have been invaluable in insulators and semiconductors. From an optical point of view semiconductors and insulators are the same, except that the band gap of semiconductors corresponds to an infrared wavelength.

Luminescence

Our main concern so far has been with the way solids transmit, absorb, scatter or reflect incident light. Sometimes they can actually emit light. The emission process is just the opposite of optical absorption and is called *luminescence*. We have seen in the previous sections how solids can absorb light when the light photons have enough energy to excite electrons or lattice vibrations in the solid. Conversely, if the electrons or lattice vibrations are excited by some other means then photons can be emitted.

The most common example of photon emission by a solid is the hot electric lamp filament, where the electrons in the metal are excited thermally and emit visible light and infrared radiation in the relaxation processes which occur. Another example, where the electrons in the solid are excited by electron irradiation rather than thermally, is the phosphor of an oscilloscope tube. In this case the conversion of energy from the electron beam to visible light is very much more efficient than in a light-bulb, where mostly infrared radiation is emitted. The phosphor used in a black-and-white television tube is zinc sulphide and can have an energy conversion efficiency of 25 per cent.

It is perhaps useful to summarize the basic features of the optical and infrared properties of ordinary solids:

(1) Dielectric solids. Here the electromagnetic wave is not absorbed or scattered (except at interfaces) but merely slowed down. Refraction is the most importance consequence of the slowing down; glass and plastics are the most common dielectric solids.

(2) Absorbing solids. These are like dielectric solids except that radiation is also absorbed, sometimes by the pure solid itself, sometimes by defects in an otherwise clear material. Coloured glasses and gemstones are examples.

(3) Diffuse scattering solids. These may be clear or coloured insulators, but so filled with bulk defects and grain boundaries that interface scattering makes them opaque. Common ceramics and bricks are good examples.

(4) Metals. Here the conduction band electrons absorb so strongly that metallic solids are always very opaque and good reflectors.

5.2. *Magnetic resonance*

Electrons and most nuclei have a small magnetic dipole moment associated with their charge spinning around on an axis, that is they

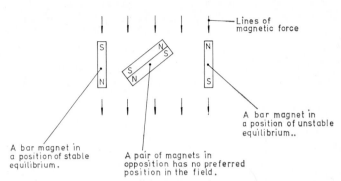

Fig. 5.10. Single and paired bar magnets in a uniform magnetic field.

behave as if they were tiny bar magnets. Just as a bar magnet has two equilibrium positions in a uniform applied magnetic field (figure 5.10) so do electrons and many nuclei. Some nuclei have no magnetic moment, but others can actually adopt more than two stable states, which serves to illustrate the inadequacy of the simple picture. However, when two states do occur one has an energy which is higher than the other by

$$\Delta E = MB \qquad (5.9)$$

where B is the flux density of the applied magnetic field and M is the magnetic dipole moment of the electron or nucleus. Then, if radiation with a frequency suitable to bridge this energy gap ($\nu = \Delta E / h$) is incident on the sample, photons can be absorbed in the process of exciting electrons or nuclei across the gap.

For electrons the characteristic frequency of the magnetic resonance absorption is 29 GHz for an applied field of 1 tesla. Protons and other nuclei have smaller magnetic moments, for protons the frequency is 42 MHz at 1 tesla and for ^{107}Ag it is only 1·9 MHz. We might guess from figure 5.10 that the radiation interacts with the electron magnetic moment (or the nuclear magnetic moment) through its own magnetic field, and that the magnetic field of the radiation must be perpendicular to the steady magnetic field (if the two fields are parallel the tiny field associated with the electromagnetic wave has essentially no effect). It turns out that the absorption of radiation in magnetic resonance is usually very small and sophisticated methods must be used to measure it, but in spite of these practical difficulties electron spin resonance (ESR) and nuclear magnetic resonance (NMR) are very widely used techniques.

NMR is most widely used in liquids, as an aid to structure determination. For example, the proton spin resonance of CH_3OH would appear as two bands, one with an intensity three times the other. In CH_3CH_2OH three bands would appear, with intensity ratios $3:2:1$, which correspond to the three different hydrogen atom environments. In solids NMR is less powerful because the width of the absorption bands is usually too great for the simple structure determination illustrated above to be possible. Nevertheless, useful information about solids can be obtained from NMR.

Ordinary insulators and semiconductors do not exhibit ESR. The reason for this is that the Pauli exclusion principle forces electrons to pair up in atoms and solids when possible, with their magnetic moments in opposition, and in this state they have no magnetic moment just as two equal bar magnets cancel when locked together (see figure 5.10). Certain special molecules like diphenyl picryl hydrazyl (DPPH) do have an odd number of electrons, one of which has to be unpaired, and they do exhibit strong ESR absorption bands, but these molecules are very rare. Defects and impurities in insulators and semiconductors, on the other hand, quite commonly have unpaired electrons and give ESR absorption. For example, the F centre in alkali halides and alkaline earth oxides has a single unpaired electron in an anion vacancy and has a strong ESR band, see figure 5.11. The structure shown in figure 5.11 arises from the way the unpaired electron interacts with nearby nuclei. This is called hyperfine structure. What happens is that the nuclear magnets usually like to line up in the magnetic field of the electron and in doing so they may reduce or increase the energy of the electron in the magnetic field. We have no time to explain hyperfine structure properly, but it is worth pointing out that it provides a very powerful tool for the determination of the structure of defects in insulators and semiconductors.

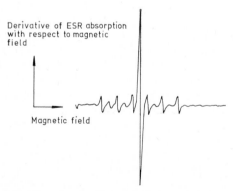

Fig. 5.11. ESR absorption due to F centres in magnesium oxide. The vertical signal is the derivative of absorption coefficient with respect to magnetic field. (after B. Henderson and J. E. Wertz, *Advances in Physics*, **17**, 749 (1968)).

88

Magnetic resonance is not involved with the way we normally experience real solids, as mechanical and optical properties are, but it does have a lot to say about the microscopic structure of solids and is, for this reason, very important.

5.3 Diffraction

Simple optical diffraction gratings are often glass plates on which lines are ruled with a spacing comparable with the wavelength of light. They transmit and reflect light of a given wavelength more strongly in some directions than others. These favoured directions, whether in reflection or transmission, are those in which the scattered waves from the rulings are all in phase. We might expect that a crystal, which consists of an ordered array of atoms, would itself act as a diffraction grating for radiation which has wavelengths comparable with the atomic spacing. We would not expect the crystal to act as a diffraction grating for optical or infra-red light, where the wavelengths are 10^3–10^4 times the atomic spacing, and certainly not for the radiation used in NMR and ESR. Fortunately, X-rays, neutrons and electrons can fairly easily be generated with wavelengths appropriate for diffraction from crystals and are all widely used for this purpose.

We can illustrate the basic principles of diffraction from crystals by looking at the square two dimensional crystal lattice in figure 5.12. Let us first look at a plane wave incident from AA′, being scattered by the two atoms at B and B′ and leaving towards C and C′. The incident wave makes an angle ϕ with respect to the line of atoms XY and the exit wave an angle ψ. If the wave at AA′ is a monochromatic plane wave the components scattered by the two atoms will be out of phase at CC′ by an amount

$$2\pi(BD-B'D')/\lambda=\frac{2\pi a}{\lambda}(\cos \psi-\cos \phi) \qquad (5.10)$$

This means that for most values of $\psi(\neq\phi)$ all the atoms in the line XY will scatter in slightly different phases, and the line will scatter as if it were composed of a number of independent jumbled atoms. If $\psi=\phi$, on the other hand, the atoms will all scatter with no phase differences and the scattered beam will be much stronger. In other words, a line of atoms (a plane in the three-dimensional case) acts like a mirror, in that the scattered beam is strong if the incident and exit angles are equal, in effect the incident beam is reflected as if by a mirror.

Let us now look at the scattering by atoms in different planes, such as atoms Q and S in figure 5.12, and make the assumption that incident and exit angles are equal (θ). Here the phase difference is

$$2\pi(PQ+QR)/\lambda=4\pi a \sin \theta/\lambda \qquad (5.11)$$

and we can make it zero only by putting $\theta=0$, which is not a very

89

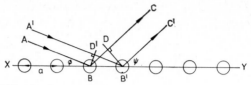

(a) Diffraction from one row

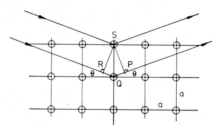

(b) Diffraction from many rows

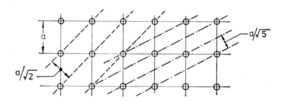

(c) Different row spacings in the lattice

Fig. 5.12. The principles of diffraction by crystals.

useful case. We can, however, make the phase difference equal to 2π, 4π, 6π and so on by choosing

$$2a \sin \theta = n\lambda \qquad (5.12)$$

where $n = 1, 2, 3 \ldots$ A phase difference of 2π, 4π and so on is in effect the same as zero-phase difference, which means that the crystal will reflect strongly if eqn. (5.12) is satisfied but not otherwise. This is the famous Bragg condition for diffraction. It allows the distance between atoms in simple structures to be determined in terms of the wavelength of the radiation used for diffraction and in more complex crystals it helps to define the way the atoms are geometrically ordered.

It is fairly clear from figure 5.12 (c) that lines or planes of atoms are to be found at several different separations. In the two-dimensional square lattice the largest separation between lines is a, but we have shown others separated by $a/\sqrt{2}$ and $a/\sqrt{5}$ and many others exist. The Bragg diffraction condition can also be satisfied for reflection from these planes but strong reflections do not always occur because the planes are not strictly equivalent as they were in the simple case.

X-ray diffraction

The wavelength of electromagnetic radiation with photon energy E is given by

$$\lambda = hc/E \qquad (5.13)$$

In this equation $hc = 1 \cdot 2398 \times 10^{-6}$ in units of eV m, so that an X-ray with energy of 10 keV has a wavelength of about 0·124 nm. The spacing between atoms in crystals is usually about 0·2–0·4 nm, so that X-rays around 10 keV have just the right size of wavelength for crystal diffraction.

The practical advantages of X-rays for this purpose are many. They are cheap and easy to produce, they pass easily through air and are readily detected with a photographic film. They are dangerous to health but shielding is so easily done that this is not a serious problem. Finally, they interact strongly with atoms, so that the diffracted beams are intense even when small samples are used. Their major drawback arises from the fact that the X-rays are scattered by the electrons in the atoms, giving a scattered intensity proportional to Z^2, where Z is the atomic number or the number of electrons per atom. This means that atoms of light elements contribute only a little to the diffraction pattern and are therefore difficult to locate. In particular it is very difficult to locate hydrogen atoms in crystals which contain heavy atoms as well.

There are several ways of carrying out X-ray diffraction experimentally. The most obvious is to use a single-crystal specimen with a monochromatic X-ray beam, and to rotate the crystal in the beam and a detector around the crystal until a diffracted beam is found. This requires sophisticated mounting equipment and two simpler ways are often used instead.

The first method, called Laue diffraction after the inventor, uses a wide range of X-ray wavelengths projected onto a single crystal with a photographic film placed around it, as in figure 5.13. As we have seen, the crystal will contain a number of different planes, each with its own spacing and orientation with respect to the incident beam. In other words, a and θ of eqn. (5.12) are fixed for each plane. However, there is a range of wavelengths in the beam and some of them will satisfy the Bragg condition, so that a diffraction pattern results (figure 5.14).

The second method, called powder diffraction, is also shown in figure 5.13. Here a single wavelength is used and the Bragg condition is satisfied because the powdered crystal has a number of crystallites with a range of orientations. Both powder and Laue methods can give the size of the unit cell in the crystal, but the single-crystal mono-

(a) Laue method

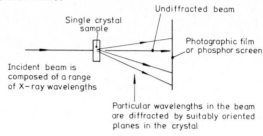

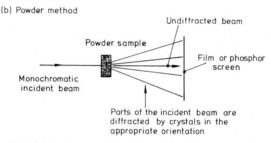

Fig. 5.13. Laue and powder diffraction techniques.

chromatic X-ray approach must be used if the arrangement of atoms within a complex unit cell is to be determined.

X-rays are scattered by the atoms in solids, and we might therefore expect that defects would affect X-ray diffraction. The most important change defects cause is of the size of the unit cell (see Chapter 4.1); the a in eqn. (5.12) changes and it can be measured using X-rays. Another effect on X-ray diffraction is through the distortions of the lattice which occur around the defects. The lattice near a defect is less perfect as a diffraction grating than before and there is more scattering at angles other than those which satisfy the Bragg condition. Both these effects are used to provide information about defects. More detailed discussion of X-ray diffraction is given in *Crystals and X-rays*.

Neutron diffraction

The wavelength of a neutron with energy E is

$$\lambda = h/(2m_n E)^{\frac{1}{2}} \tag{5.14}$$

where m_n is the neutron mass. This means that a wavelength of 0·2 nm is achieved when the neutron energy is about 0·025 eV. Fortunately for neutron crystallographers this is the energy neutrons reach if they are allowed to diffuse through a solid at room temperature, and many of the neutrons in nuclear reactors also have energies around 0·025 eV.

Fig. 5.14. A Laue diffraction pattern caused by uranium oxide. In this case a four-fold symmetry axis of the cubic UO_2 crystal is aligned along the incident X-ray beam (courtesy of C. Sampson).

Reactors are by far the most powerful sources of neutrons with the right energies for diffraction by crystals. They are uncomfortably more expensive than X-rays, a sophisticated X-ray diffraction instrument might cost £10 000, whereas a good neutron beam reactor will be more than £10 000 000, but they are nevertheless used. Neutron diffraction has also been dealt with in this series, in *Neutron Physics.*

Neutrons have two important advantages over X-rays for some

93

diffraction experiments. First the scattering of hydrogen atoms is very strong, and neutrons can readily locate light atoms in the presence of heavy ones. Secondly, neutrons have a magnetic moment and can measure the magnetic structure of crystals such as iron as well as their geometric structure. Apart from their expense, neutrons have the disadvantages of requiring large specimens and of being difficult to detect and to shield.

One of the most interesting ways in which they can be used to study defects is in long-wavelength neutron scattering. If the neutron wavelength is long enough it is not possible to satisfy the Bragg condition of eqn. (5.12) for any value of θ, and we would expect the neutrons to pass straight through the crystal. Under these conditions a crystal containing fairly large defects will scatter neutrons very differently from a perfect one, the scattering difference being due to the defects themselves and providing direct information about the defects. Although in principle this can also be said of X-rays, in practice it cannot be done because X-rays interact with the atoms of the solid in too many other ways, which will be discussed in Chapter 6.

Low energy electron diffraction

The wavelength of an electron with energy E is also given by eqn. (5.14), but with the electron mass replacing the neutron mass. Since neutrons at 0·025 eV have a wavelength of 0·2 nm it follows that electrons at about 2000×0.025 eV $= 50$ eV will have the same wavelength. Electrons at 50 eV can easily be produced and are therefore used for low energy electron diffraction (LEED).

The major difference between 10 keV X-rays, thermal neutrons and 50 eV electrons is in their penetration into solids. The X-rays will typically penetrate a few hundred μm, the neutrons several tens of mm but the electrons will only penetrate a few atomic layers. For this reason LEED is useful in determining the structure of surface layers, such as the Al_2O_3 corrosion layer on Al, but is quite useless for bulk measurements. The experimental facilities required for LEED are not very expensive but neither are they trivial, since very highly evacuated equipment must be used to prevent the relatively slow electrons being scattered by occasional atoms in the gas.

Electron microscopy

Low energy (50 eV) electrons have wavelengths comparable with the distance between atoms in crystals, but they are difficult to use. Electrons with energies of about 100–200 keV are much easier to manipulate but have wavelengths of about 0·04 nm. Nevertheless, they can still be diffracted by crystals, the Bragg condition can be satisfied by having n larger or by using planes which are close together in the crystal, that is, small a.

Simple high energy electron diffraction is not of great importance, though it does serve as a good example of the duality of particles and waves (see *Elementary Quantum Mechanics*). However, when coupled with electron imaging it provides the basis of an enormously powerful technique for studying large defects such as dislocations.

Let us look at the diagram of an edge dislocation shown in figure 5.15, at an angle which nearly but not quite satisfies the Bragg condition. As shown by the set of arrows on the right and left the crystal well away from the core of the dislocation is effectively perfect, and the intensities of the transmitted and refracted beams will be constant. Nearer the core of the dislocation the crystal planes are increasingly bent, with the result that there the crystal is locally no longer in the same orientation as far away from the dislocation. In the schematic drawing of figure 5.15

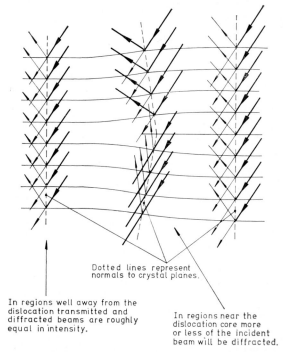

Dotted lines represent normals to crystal planes.

In regions well away from the dislocation transmitted and diffracted beams are roughly equal in intensity.

In regions near the dislocation core more or less of the incident beam will be diffracted.

Fig. 5.15. Electron diffraction near a dislocation.

planes above the core are bent nearer the Bragg condition and diffract more strongly, while those below the core are bent away from the Bragg condition and diffract less strongly. Because of this the distorted region near the dislocation core can be distinguished from perfect crystal by imaging either the transmitted or diffracted electrons with a suitable electron lens on to a phosphor screen or photographic film as shown in figure 5.16.

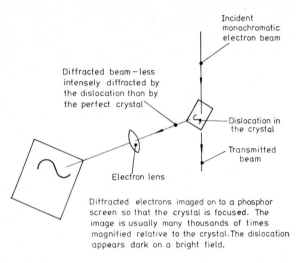

Incident monochromatic electron beam

Diffracted beam – less intensely diffracted by the dislocation than by the perfect crystal

Dislocation in the crystal

Transmitted beam

Electron lens

Diffracted electrons imaged on to a phosphor screen so that the crystal is focused. The image is usually many thousands of times magnified relative to the crystal. The dislocation appears dark on a bright field.

Fig. 5.16. A schematic representation of the electron microscopy of a dislocation.

The defect image may be darker or lighter than the background intensity from the surrounding perfect crystal, depending on the orientation chosen for the crystal in the electron beam and whether transmitted or diffracted electrons are used for imaging. An electron micrograph of a dislocation in copper is shown in figure 5.17. Other defects, such as stacking faults, voids or precipitates, can also produce contrast in the electron microscope because they scatter electrons in a different way from the perfect crystal.

In this chapter we have looked at how various kinds of radiation interact with solids and with defects in solids. Always we have assumed that the radiation leaves the solid as it finds it, that no atoms are moved permanently or electrons removed. In this sense we have been dealing with low energy radiation. In general, optical and infrared photons cannot change solids permanently, but X-rays and high energy electrons and neutrons can. Before we find out how this is done we need to look more carefully at how high energy photons and particles transfer energy to solids, for it is this energy transfer which causes the 'radiation damage'.

96

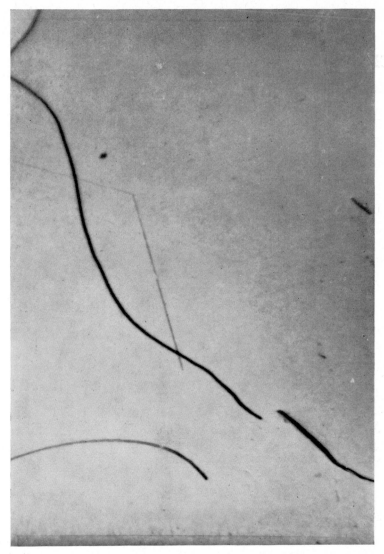

Fig. 5.17. An electron microscope photograph of a dislocation in copper (courtesy of L. W. Hobbs).

CHAPTER 6

interaction of solids with high energy radiation

There is really no well-defined demarcation between high energy and low energy radiation, although for our purposes it is convenient to think of high energy radiation as that which can cause permanent changes to the solid and low energy radiation as that which cannot. This classification is not without its disadvantages because the same radiation can cause permanent changes in some solids and not in others. For example, the kind of X-rays which are used in medical radiography have no permanent effect whatever on metals but cause organic chemicals to decompose. Nevertheless, the guide that high energy radiation causes permanent changes, which are usually called radiation damage, is one which we shall use.

From the very beginning of work on high energy radiation from natural radioisotopes it has been known that both electromagnetic radiation and fast atomic particles are important sub-groups. Now, with our greater understanding of radiation effects and with the great variety of man-made radiation sources at our disposal, we can more usefully classify high energy radiation into the following four sub-groups.

1. High energy photons (X-rays, γ-rays).
2. Light charged particles, such as electrons, protons, α particles.
3. Heavy atoms or ions such as argon.
4. Neutrons.

We will look at each group in turn, concentrating particularly on how they interact with solids.

6.1. *High energy photons*

The photon is the quantum of electromagnetic radiation. Its energy E is related to the frequency (ν) and the wavelength (λ) of the radiation by eqn. (5.3).

When 'high energy' is used to describe photons we normally imply that these energies are larger than a few hundred electron volts. However, the important energy range is 10 keV to 10 MeV, corresponding to a wavelength range from 0·12 nm to 0·12 pm, since in this range most X-ray generators and isotope sources have the bulk of their output. Incidentally, there is no real difference between X-rays and γ-rays;

people usually call radiation from electric generators X-rays and that from isotope sources γ-rays, but by no means invariably.

Photons in the energy range 10 keV to 10 MeV interact with solids in three principal ways, by *absorption, scattering* and *pair production*. In the first case the photon is completely absorbed by an atom in the solid; in the second only part of the photon energy is absorbed, the rest being carried away in a scattered photon of lower energy than the original. In pair production the photon is absorbed by the solid and creates new atomic particles (electron–positron pairs) in the process.

Photo-electric absorption

This is a very straightforward process. As we have seen in earlier chapters, the electrons in a solid exist in well-defined energy bands with varying binding energies, and the bands in which electrons are most tightly bound are full. Above these full bands there are many empty electron states, to which electrons from lower levels can readily be excited by an incident photon. Provided the incident photon has enough energy to excite electrons in this way, it will have a certain probability of being absorbed, and where absorption occurs its energy is used entirely in creating a high energy electron and a hole in a deep-lying energy band in the solid. When atoms near the surface of the solid are excited the same process occurs, but then the electron may be given enough energy to leave the solid altogether. Hence the process is often called photo-electric absorption and the ejected electron is called a photo-electron.

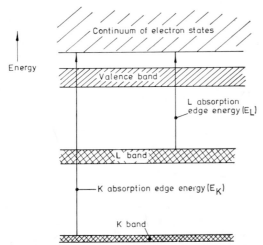

Fig. 6.1. The energy-band description of photo-electric absorption in solids.

Suppose we have a solid with energy bands as shown in figure 6.1 and we irradiate it with high energy photons. The most tightly bound

99

electrons in atoms are in the K shell, and in solids are said to be in the K band. In the same way the higher bands are called L, M, N and so on. Once the energy of an incident photon is larger than E_L it will be possible to excite electrons from the L band into the continuum of empty states; at energies less than E_L such excitation will not be possible and E_L is therefore called the L absorption edge. As the photon energy is increased beyond E_L its absorption excites L-band electrons to higher and higher energy states in the continuum, but with steadily decreasing probability. The result is that the absorption co-efficient falls steadily with increasing photon energy. However, once the photon energy reaches E_K, the K-band absorption edge, the absorption coefficient rises sharply because it becomes possible to excite not only L but also K electrons. The absorption coefficient therefore has a sawtooth appearance as a function of photon energy as shown in figure 6.2.

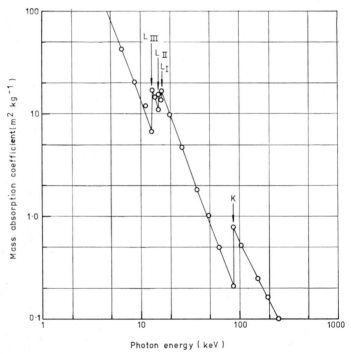

Fig. 6.2. Photo-electric absorption coefficient of lead as a function of energy.

We can define a photo-electric absorption coefficient for this process. As before, if the initial photon-beam intensity is I_o, after passing through a thickness x it will be reduced to

$$I(x) = I_o \exp(-\alpha x)$$

100

where α is the macroscopic absorption coefficient of the material and usually has values of around 10^3–10^5 m^{-1} for photons in the energy range 10 keV to 100 keV. The atomic absorption coefficient σ is just α divided by the number of atoms per unit volume in the solid, and is a very useful concept because it allows absorption coefficients for compounds or mixtures of atoms to be calculated from those for the constituent atoms. Notice, by the way, that σ has the dimensions of a cross-sectional area. For all except light atoms it has been found empirically that the atomic absorption coefficient, or the atomic absorption cross-section as it is sometimes called, is given by

$$\sigma = C\, Z^4 / E^3 \qquad (6.1)$$

for absorption of a photon of energy E by an atom with atomic number Z. The constant C depends on whether E is above the K edge ($C_{K+} = 2 \times 10^{-18}$ m^2 eV^3) or between the K and L edges ($C_{KL} = 3 \times 10^{-19}$ m^2 eV^3) and so on, but it does not depend on the kind of atom absorbing. The quantum-mechanical derivation of eqn. (6.1) is too complex for us to give here but the general dependence on E and Z can be understood qualitatively as follows. In the absorption process momentum can be conserved only if some is given to the atom as a whole as well as to the excited electron, and this can be done only through Coulomb interaction between electrons and nucleus. For larger Z a stronger interaction with the nucleus is possible, hence the absorption coefficient increases with increasing Z. For larger E much more momentum must be transferred to the nucleus, and since this is difficult the absorption coefficient falls as E increases.

It is clear from eqn. (6.1) that photo-electric absorption becomes unimportant for light atoms and for high energy photons. Usually it dominates the interaction between photons and matter in the 10–100 keV range and is negligible above 1 MeV. In the intermediate range Compton scattering often dominates, particularly for light elements, so we will now look at that.

Thomson and Compton scattering

If a photon interacts with a particle or an atom and is scattered by it, energy can be given to the particle, the photon leaving the collision with a smaller energy than before. If the energy of the photon is small enough it can interact with the atom as a whole, and is then scattered with very little loss of energy. The maximum energy which can be lost to an atom in this way can be calculated very easily, provided we remember that the momentum of a photon of energy E is E/c, where c is the velocity of light. The maximum energy transfer occurs when the photon is scattered backwards as in figure 6.3.

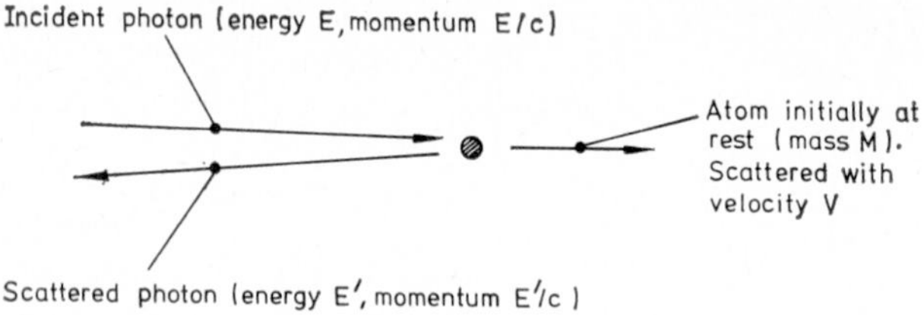

Fig. 6.3. The back scattering of X-ray photons by an atom.

The law of momentum conservation tells us that

$$E/c + E'/c = MV \qquad (6.2)$$

and that of energy conservation gives

$$E - E' = \tfrac{1}{2}MV^2 \qquad (6.3)$$

These equations can be solved for V, and hence $\tfrac{1}{2}MV^2$, in the normal way, but we can approximate quickly since the energy transfer is small, and so $E \sim E'$. Equation (6.2) gives, with this approximation,

$$V \simeq 2E/Mc$$

It follows that the energy transferred to the atom is

$$\tfrac{1}{2}MV^2 = 2E^2/Mc^2 \qquad (6.4)$$

Since the term Mc^2 is 1000 MeV even for a hydrogen atom, it is clear we were right to assume small energy transfer for all reasonable photon energies. This Thomson or atomic scattering process is often called elastic scattering because the scattered photon has lost so very little energy. It becomes very unlikely when E is large because it requires that the photon first interact with the electrons in the atom and that these in turn interact with the nucleus. For high photon energies it becomes more probable that the photon will tear one of the electrons off the atom as the interaction occurs. For example, the probability of elastic scattering of keV photons by carbon atoms is negligible, yet even under these conditions the maximum energy transfer is only 0·06 eV. For this reason Thomson scattering is not important as a process by which radiation can lose energy, nor does it contribute at all to radiation damage. It is, however, the physical basis of the very important phenomenon of X-ray diffraction, which we discussed in Chapter 5 without specifying how the atoms interacted with the photons.

If the energy of the photon is high it will be scattered by one of the electrons in the solid as if it were a free electron unattached to its parent nucleus. In this case, which is called Compton scattering, the maximum energy transferred to the electron must be calculated relativistically and is given by

$$2E^2/(2E+m_ec^2) \tag{6.5}$$

where m_e is the electron rest mass. The maximum energy transferred to an electron by a 1 MeV photon is about 0·8 MeV, in which case the scattered photon would have an energy of 0·2 MeV.

The probability that a Compton scattering event will occur is sometimes expressed in terms of an attenuation coefficient α for the solid, exactly analogous to the absorption coefficients we have used previously, or alternatively, as a scattering cross-section per electron, written σ_e and analogous to the absorption cross-section per atom we used in the last section. If the number of electrons per unit volume in the solid is N_e m^{-3}, then

$$\alpha = N_e\sigma_e \text{ m}^{-1}$$

The intensity of unscattered photons after passage through a thickness x of the solid is then given by

$$I(x) = I_o \exp(-\alpha x)$$

as before.

For Compton scattering of low energy photons the cross-section per electron is given by

$$\sigma_e = \sigma_{eo} = \frac{e^4}{6\pi\epsilon_o^2 m_e^2 c^4} = 6\cdot7 \times 10^{-29} \text{ m}^2 \tag{6.6}$$

At high photon energies E the cross-section per electron is given by:

$$\sigma_e = \sigma_{eo} \frac{3}{8E'} (\tfrac{1}{2} + 2\ln 2E') \tag{6.7}$$

where $E' = E/m_ec^2$. This falls off much less rapidly at high energies than does photo-electric absorption and for that reason it often dominates photon–atom interactions in the 100 keV to 1 MeV range. Figure 6.4 plots the cross-sections per atom for photo-electric absorption and for Compton scattering for carbon and copper atoms as a function of photon energy. It shows how photo-electric absorption dominates at lower photon energies but falls below Compton scattering in importance as the photon energy increases. It is worth emphasizing that eqns. (6.6) and (6.7) give the cross-section for Compton scattering per electron. For an atom with atomic number Z there are Z electrons so that the Compton cross-section per atom varies linearly with Z. As we saw in eqn. (6.1) the photo-electric absorption per atom varies with Z^4, which means that for light atoms Compton scattering is important at lower energies than for heavy atoms. The direct products of Compton scattering are high energy electrons, their associated holes, and scattered photons with reduced energy. The scattered electrons are most likely to

103

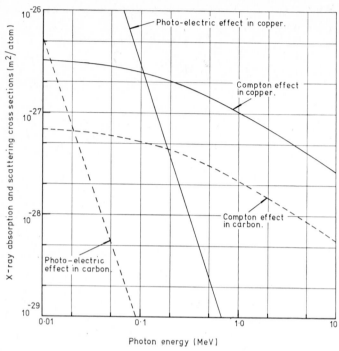

Fig. 6.4. Photo-electric and Compton scattering cross-sections for carbon and copper atoms, as a function of energy.

come from loosely bound energy bands so that the energy associated with the remaining hole is not usually very significant.

Pair-production

Once the high energy photon has energy in excess of $2m_ec^2$ ($=1\cdot022\,\text{MeV}$) it can create electron–positron pairs. The energy E of the photon is used directly in the creation of the mass of two new particles, hence E must be larger than $2m_ec^2$. The nucleus of an atom is required once again for conserving momentum and causes the probability of a pair-production event to vary with Z^2. The probability rises fairly rapidly with photon energy as this rises from $2m_ec^2$. At $E=2m_ec^2$ it is clear that a pair-production event can create only electrons and positrons at rest, whereas at higher energies a range of possible electron and positron momenta are allowed, making the probability higher. We often express this by saying that the density of final states is larger at higher photon energies.

In our discussions so far, of photo-electric absorption, Compton scattering and pair-production, we have said nothing of what causes the high energy photon to interact with the atoms in each case. For photo-electric absorption and Compton scattering the interaction is just through the electric field component of the electromagnetic radiation acting on the negatively charged electrons, in much the same way as the electric field component of a radio wave acts on the electrons in a metal aerial and causes them to move about. The pair-production interaction is not quite so easy to explain. How can the field of the electromagnetic radiation act on particles which do not yet exist? This is one of the problems of quantum electrodynamics and far beyond the scope of our discussion, but it is worth mentioning one of the techniques used in quantum electrodynamics to describe events like pair-production. This considers the positron leaving the pair-production event to be an electron moving backwards in time, and this proves to be a very helpful concept. From our point of view it at least gives us an electron with which the photon can interact.

Cross-sections

As we have said often already, the intensity of a beam of photons passing through a solid falls exponentially with the thickness of material traversed. It is therefore very common to use the constant α, which is the attenuation or absorption coefficient of the *material,* to describe the strength of the interaction between photon and solid.

The use of an attenuation coefficient for each material is necessary for low energy photons but less desirable for high energy ones, where a given *atom* contribution to α does not depend very much on just how it is incorporated in the solid. To illustrate this difference between low and high energy photons let us consider two hypothetical structures; the first a layer of crystalline silicon together with a layer of solid oxygen and second a layer of SiO_2 containing exactly the same amounts of Si and O_2 as the first structure. As far as optical transmission is concerned the two structures are totally different. Crystalline silicon is completely opaque to all visible and ultraviolet light and even solid oxygen absorbs ultraviolet light strongly at wavelengths below 200 nm. Yet the SiO_2, containing the same atoms, is completely transparent to all visible and ultraviolet photons, right down to wavelengths of 170 nm. Under these conditions it is clearly unhelpful to talk about the contribution to α of the Si or O atoms, since this contribution depends so strongly on the exact state of these atoms. The reason is that the optical properties of solids are determined by the valence electrons, and their character depends strongly on just how the atoms are bound together. In contrast the absorption of X-rays depends on tightly bound electrons not greatly influenced by the character of bonding. Because of this, at high photon energies, from a few keV upwards, the attenuation of the two structures would be the same, and it turns out that α for any material can be expressed as

$$\alpha = \sigma_1 N_1 + \sigma_2 N_2 + \ldots \tag{6.8}$$

where $N_1, N_2 \ldots$ are the numbers of constituent atoms per unit volume of type $1, 2 \ldots$ in the material and $\sigma_1, \sigma_2 \ldots$ are the atomic absorption coefficients, or the cross-sections for interaction, with each atom. We have already pointed out that the σ's must have the dimensions of area or cross-section, since α has the dimensions of *length*$^{-1}$ and the atomic densities N have *length*$^{-3}$. Thus the concept of cross-section allows considerable simplification in determining α; once we know σ as a function of energy for a mere $90+$ elements we can work out α for many hundreds of thousands of compounds and mixtures.

We have arrived at the cross-section concept via absorption coefficients. If we had begun our discussion with particles rather than photons we would have reached the same conclusion another way, a point which seems worth mentioning now.

Suppose, as in figure 6.5, we have a table top of area A, on which are sitting $\mathcal{N}$ tennis balls each of radius r and cross-sectional area $\sigma(=\pi r^2)$. If a very small ball bearing is dropped onto the table from a random position over it, as in figure 6.5, then the probability that the ball bearing will hit a tennis ball rather than the table is $\mathcal{N}\sigma/A$. In just the same way the probability that a photon, or for that matter a particle,

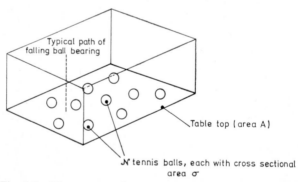

Fig. 6.5. The concept of a cross-section in particle scattering.

will interact with an atom in passing through a layer of atoms of often expressed as $\mathcal{N}\sigma/A$ where $\mathcal{N}/A$ is the number of atoms per unit area in the layer and σ is the atomic cross-section. If there are n such layers in a thickness x of material the total probability of an interaction with the photon or particle is

$$n\mathcal{N}\sigma/A = \left(\frac{n\mathcal{N}}{xA}\right)\sigma x$$

106

Now $n \mathcal{N}/xA$ is just the number of atoms per unit volume in the solid (N_o) so that the probability that a photon or particle will interact with an atom in a layer of thickness x is just

$$p = N_o \sigma x \qquad (6.9)$$

The cross-section σ gives the total probability that a collision between a photon or particle and an atom will occur, and can be thought of as the effective total area the atom presents to the radiation.

Collisions between a small ball bearing and a tennis ball can obviously vary a great deal in character, from a violent, head-on collision to a glancing one which scarcely affects either object. The same is true of collisions between particles or photons and atoms. For this reason it is usual to sub-divide the kinds of collision, either in terms of the angle through which the particle is scattered or, probably more useful to us, the energy transferred to the target atom. The probability that a given kind of collision will occur is given by an equation just like (6.9) but with the total cross-section σ replaced by a *differential cross-section* $K(E)$. For example, the probability that a photon would, in a Compton scattering event, transfer an energy between E and $E+dE$ to an electron in an atom is usually written

$$p(E) = N_o \, K(E) \, dE \, x \qquad (6.10)$$

It can be seen that the integral of the cross-section for transferring an energy E, over all energies up to E_m, the maximum which can be transferred, must equal the total collision cross-section. That is

$$\sigma = \int_0^{E_m} K(E) \, dE \qquad (6.11)$$

The generation of high energy photons

After this digression into cross-sections, we should now look briefly at how the high energy photons we have been discussing are generated. The commonest natural sources are radioactive elements. Just as atoms have well-defined energy levels for electrons, and those atoms with excited electrons emit light of well-defined frequency, so nuclei have energy levels for neutrons and protons and, if excited, they too can emit electromagnetic radiation. The primary difference is that atoms emit visible or ultraviolet light, or X-rays in the case of very heavy atoms, whereas excited nuclei usually emit very high energy photons ($\geqslant 100$ keV). The lifetime of excited atoms and nuclei is always very short, but they are formed steadily in nature when radioactive nuclei emit α-particles (helium atom nuclei) or β-particles (electrons) which they do slowly and steadily. In most cases the new nuclei which are formed are born in an excited state and emit γ-ray photons in decaying to their ground state.

Fig. 6.6. The spent fuel pond γ-irradiation facility at Harwell (courtesy of the U.K.A.E.A.).

Man-made radioactive nuclei, such as ^{60}Co and ^{137}Cs, are now more widely used than natural ones as convenient γ-ray sources. These two emit photons of energy 1·2 MeV and 0·67 MeV respectively. Figure 6.6 shows a γ-ray facility at Harwell. Here the γ-rays originate from the variety of radioactive elements in 'spent' fuel rods from nuclear reactors. The visible glow is Cerenkov radiation, caused by the Compton scattered electrons moving faster than the speed of light in the water used for shielding.

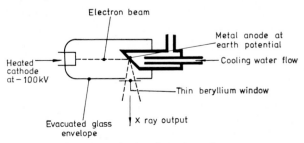

Fig. 6.7. A schematic drawing of an X-ray generator.

Another important source of high energy photons is the X-ray generator, shown schematically in figure 6.7. These machines are high vacuum thermonic diode valves, similar in principle to those used in pre-transistor radio receivers but very different in practice. The heated cathode is maintained at a large negative voltage, usually up to 100 kV, and emits electrons which are accelerated to an anode which is at earth potential and is water-cooled because of the high electron beam power dissipated in it. When the electrons hit the metal target some of their energy is converted into X-ray photons because the incoming electrons are decelerated so rapidly by the heavy nuclei in the target. The X-radiation which results is often called 'bremsstrahlung' which is German for braking-radiation. The X-rays which are produced in bremsstrah-

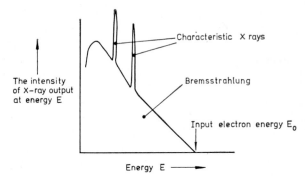

Fig. 6.8. The energy distribution of bremsstrahlung and characteristic X-radiation.

lung are not monochromatic but are distributed in energy between zero and the energy of the incident electron as shown in figure 6.8. Bremsstrahlung is the important type of radiation for use in X-radiography and radiation therapy.

Figure 6.8 indicates that some monochromatic radiation is also generated in X-ray tubes of the kind shown in figure 6.7. This is called characteristic radiation because its wavelength depends on the atoms of which the target is made and not on the incident electron beam energy. It occurs because of electronic transitions between the deep energy levels in the atoms of the target, following the excitation of the target atoms by the electron beam.

6.2. *Light charged particles*

We will deliberately limit this section to a discussion of the way light charged particles interact with matter because we want to avoid having to worry about whether an ion traverses the solid as a bare nucleus or as an ion which includes a large number of electrons. In the latter case ions often interact with other atoms as if they were of considerable size, and quite different phenomena result. High energy protons, α-particles and electrons are the best examples of light charged particles; as long as protons and α-particles have more than a few keV of energy they pass through matter as H^+ and He^{2+} respectively, so we need never be in any doubt with what we are dealing.

Energy loss to electrons and nuclei

The most important mechanism for interaction between these particles and the atoms in solids is coulombic. That is, the electrostatic forces between the incoming particle and the electrons and nuclei in the solid cause energy transfer collisions to occur. The first question which comes to mind is how much energy is lost to the electrons of the solid compared with that lost to the nuclei, and we can fortunately answer it with the aid of very simple arguments.

Suppose a high energy proton or electron passes an atom in a time t. We can talk of the particle and atom interacting for a definite time t for two reasons. First, the electrostatic force between the particle and the electrons and nuclei in the atom falls with the distance like R^{-2}, so that for large separations they essentially do not interact at all. Secondly, in a solid even this force is shielded by the electrons in other atoms if these are between the two interacting particles. As a result significant interaction occurs only over a certain distance s (of the order of the size of the atom) and for a time t. During this time the average force exerted by the particle on a single electron in the atom will be F and that on the nucleus ZF because of its higher charge. It follows that the velocity imparted to an electron in the solid by the passing particle is

$$u=Ft/m_e$$

so that it would gain energy

$$\tfrac{1}{2}m_e u^2 = \tfrac{1}{2}(Ft)^2/m_e \tag{6.12}$$

The nucleus of mass M, on the other hand, would gain a velocity

$$U=ZFt/M$$

and energy

$$\tfrac{1}{2}MU^2 = \tfrac{1}{2}Z^2(Ft)^2/M \tag{6.13}$$

Since the atom has only one nucleus and Z electrons we would expect the ratio (energy transferred to electrons: energy transferred to atoms) to be

$$[Z\tfrac{1}{2}(Ft)^2/m_e]/[\tfrac{1}{2}Z^2(Ft)^2/M]=M/Zm_e$$

This ratio is between 2000 and 6000; that is, far more energy will be lost to electrons than directly to nuclei.

The importance of what we have just uncovered, that high energy charged particles lose almost all of their energy to the electrons of a solid and only a tiny fraction directly to nuclei, can scarcely be over-stated. We shall see in Chapter 7 that only energy transferred directly to nuclei can contribute to radiation damage in many materials, which means that in these materials radiation damage is relatively inefficient. On the other hand, those materials where electron excitation can lead to radiation damage are damaged much more readily.

Energy loss by electrons and heavier particles

Having seen how a given particle transfers energy to electrons and nuclei let us now compare different particles. Again we can use simple arguments. In eqn. (6.12) we wrote down the energy loss of the particle to each electron in the solid so that the energy loss per atom is proportional to:

$$Z(Ft)^2/m_e$$

Let us assume the particle has charge ze. The force (F) between it and each electron in the atom will then be proportional to z. If the particle has velocity v the time (t) taken to pass the atom will be proportional to $1/v$, so that the energy loss per atom will be proportional to

$$\Delta E \propto Zz^2/m_e v^2 \tag{6.14}$$

Very much more sophisticated calculations give a dependence of energy loss on Z, z and v which is not very different from that given in eqn. (6.14).

111

If the energy loss in a single collision is as given in eqn. (6.14), the total energy loss in traversing a thickness dx will be proportional to:

probability of colliding with an atom $\times$ energy loss per collision
$= N_o \, \sigma \, dx$ (from eqn. (6.9)) $\times Zz^2/m_e v^2$ (from eqn. (6.14))

The cross-section in this model is essentially the cross-sectional area of the atom, and does not vary very much with z, so we have

$$dE \propto N_o Zz^2 dx/v^2$$

Since $E = \frac{1}{2}mv^2$, where m is the particle mass and v its velocity, we can write

$$\frac{dE}{dx} \propto N_o Zz^2 m/E \tag{6.15}$$

If we integrate this equation between $E = E_o$ at $x = 0$ and $E = 0$ at $x = R$, where R is the particle range, we have

$$R \propto E^2/N_o Zz^2 m \tag{6.16}$$

Thus we have shown that the range of a given particle should be proportional to the square of its energy; this is found to be the case for electrons, protons and other particles provided their energy is high enough. We have also shown that the range should be inversely proportional to $N_o Z$ which is the number of *electrons* per unit volume in the target material, (N_e) and again this is found to be the case. Finally we have derived the dependence of the range on the properties of the projectile, through z and m. This tells us that protons should have 16 times the range of α-particles of the same energy, since they have one-half of the charge and one-quarter of the mass, and twice the range of deuterons, since protons have half the mass and the same charge as deuterons. The dependence on $1/m$ suggests that electrons should lose energy much more slowly and have a much larger range than protons and heavier particles. This is the case, although the range of 1 MeV electrons is only about 100 times that of protons of the same energy rather than 2000 times as expected from eqn. (6.16). The primary reason for this is the relativistic nature of 1 MeV electrons. Their velocity is nearly constant rather than depending on $E^{\frac{1}{2}}$. Nevertheless, our very simple analysis has highlighted all the basic features of the ranges of charged particles.

Scattering and straggling

So far we have ignored the subtler effects of a collision between a charged particle and an atom. One important factor is the angle through which the particle itself is scattered by the collision. Even from the simple impulse approximation that we have used we can argue that these changes in *direction* are caused mainly by interaction with the atomic *nuclei*, in contrast to the *energy losses* which occur via the atomic *electrons*. Remember that the force between particle and nucleus was proportional to Zz, leading to a transverse velocity of the scattered particle proportional to Zzt, whereas in the same atom there were Z interactions with electrons and in each the force is proportional to z. The changes in direction caused by the Z electrons would be random and tend to cancel, to give a transverse velocity proportional to $zt\sqrt{Z}$, so that only the nuclear contribution need be considered provided Z is not too small.

In most collisions between a particle and an atom in a solid, even the nuclear contribution is very small. Nevertheless, a particle which passes through many atomic layers will ultimately be scattered through a significant angle. The force between particle and nucleus is proportional to Zz (remember that ze is the charge of the particle and Ze that of the target nucleus) so that the transverse acceleration of the particle will be proportional to Zz/m, where m is the particle mass. The interaction time t is proportional to $1/v$ and hence the change in transverse particle velocity due to the collision is proportional to Zz/mv. There will be equivalent changes in longitudinal particle velocity too, but they can be ignored because the initial longitudinal velocity is high, whereas the initial transverse velocity is assumed to be zero. This situation is illustrated in figure 6.9, which shows that the average scattering angle for one collision is

$$\theta \propto Zz/mv^2 \qquad (6.17)$$

If the particle passes through a thickness x of material which has density ρ, then the number of layers traversed will be proportional to

$$n \propto \rho x \qquad (6.18)$$

The scattering in the various layers will be unrelated in direction, which

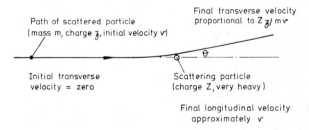

Fig. 6.9. Small-angle scattering of a fast particle by a heavy nucleus.

113

means that the final average scattering angle ϕ will be related to θ and n by

$$\phi \propto \theta \sqrt{n} \propto \frac{Zz}{mv^2}(\rho x)^{\frac{1}{2}} \qquad (6.19)$$

This equation gives the correct dependence on Z, z, m, etc. The process is called multiple coulomb scattering, for fairly obvious reasons.

In spite of multiple scattering, protons and α-particles are not deviated very much by the atoms in a solid. They travel in a fairly straight line until near the end of their tracks. Equation (6.19) suggests that, for a given particle energy (mv^2 constant), electrons will be no more strongly scattered than protons. They are more strongly deviated, partly because they can be scattered through large angles by collisions with other electrons whereas the heavier particles cannot, and partly because, for a given energy they travel much further, allowing x and therefore ϕ in eqn. (6.19) to take on larger values. This behaviour is illustrated in figure 6.10. One of its most important consequences is that a layer of material which is thinner than the range of the electrons incident on it, and ought therefore to pass all the electrons, will in fact stop a significant fraction because their path through the layer is so tortuous. Under the same conditions almost all the protons would

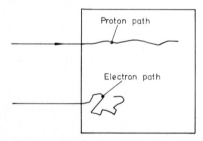

Fig. 6.10. Typical paths of high energy protons and electrons in solids.

pass successfully through the foil. The fraction of incident particles which do pass through a foil of a given thickness is shown in figure 6.11. In this figure the electrons and protons have been assumed to have the same range, although the protons would have to have a much higher energy for this to be so.

One final effect in the passage of light charged ions through solids is worth mentioning. In our previous discussions on energy loss we talked about the average energy loss per atom, for example, eqn. (6.14). As a result we implied that, provided the particle followed a fairly straight path, it would have a well-defined range. In fact some particles will lose more energy than average per collision and have shorter ranges, others will consistently lose less and have longer ranges.

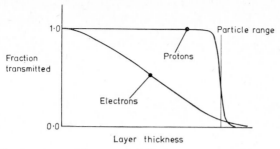

Fig. 6.11. The fractions of beams of high energy electrons and protons which pass through a solid.

This phenomenon is called range straggling. Strictly it must be considered in describing the passage of charged particles through solids, but in practice it is not usually very important.

6.3. *Heavy ions*

Let us now consider what happens when a heavy ion, consisting of a nucleus and a number of attached electrons, passes through a solid. How does its interaction differ from that of the bare nucleus or electron which we discussed in the last section? There prove to be two important aspects of this problem. First we need to find out under what conditions the moving particle can ionize the atoms in the solid or itself be ionized in passing through the solid, and secondly we need to explore the character and strength of genuine atom–atom interactions.

Energy limits for stripping and ionization

In the last section we showed that high energy charged particles are likely to lose most of their energy to the electrons of the solid through which they are passing. However, we assumed in eqn. (6.12) that an arbitrarily small amount of energy could be transferred to the electrons of the solid if necessary. In practice the energy loss to electrons is negligible unless an electron can be given sufficient energy for it to leave the atom. It follows that an ion travelling slowly through a solid may not be able to lose energy to electrons, because the energy it can transfer to an electron may not be large enough to ionize an atom.

Suppose we have an ion of mass m passing at velocity v through a solid. If it collides head-on with an electron in the solid the latter will be knocked forwards with a velocity u, as shown in figure 6.12, and the ion will be slowed to a velocity v'. In order to conserve momentum and energy we must have

$$m(v-v')=m_e u \qquad (6.20)$$

$$\tfrac{1}{2}m(v^2-v'^2)=\tfrac{1}{2}m_e u^2 \qquad (6.21)$$

115

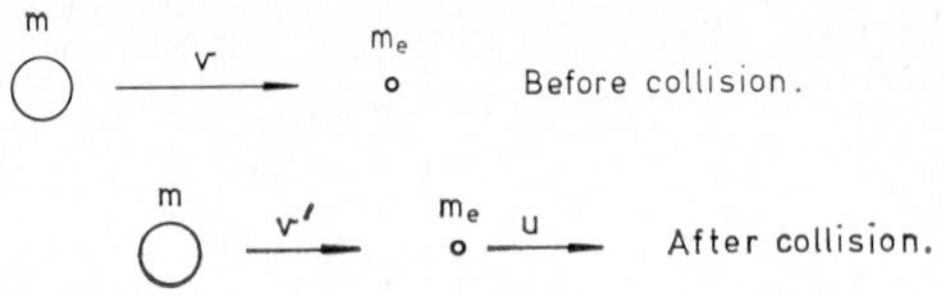

Fig. 6.12. A collision between a heavy particle and an electron in a solid.

If the ion is heavy its velocity will not change much in the collision, so that we can solve eqns. (6.20) and (6.21) by dividing (6.21) by (6.20) and making the approximation $v+v'\sim 2v$. This gives $u=2v$ so that the energy transferred to the electron is

$$\tfrac{1}{2}m_eu^2=\frac{4m_e}{m}\tfrac{1}{2}mv^2=4(m_e/m)E \tag{6.22}$$

where E is the kinetic energy of the incident ion. If this energy which is transferred to the electron does not exceed some representative ionization energy for the atoms in the solid, then energy loss to electrons is not possible.

A fairly good approximation is to say that if the ionization potential of the target atom is I_t then the projectile energy must exceed

$$E_I=I_tm/4m_e \tag{6.23}$$

For most atoms $I_t\sim 5$–10 eV, so that for proton projectiles the ionization threshold is 2–5 keV, whereas for much heavier argon ion projectiles the threshold will be around 100–200 keV. Equation (6.23) explains our limiting the discussion of Section 6.2 to *light* charged particles. Light charged particles do lose energy predominantly to electrons over almost the whole of their range but heavy ions do not, because the ionization threshold E_I is so large for heavy ions, and below E_I little energy loss to electrons can occur.

What we have said about removal of electrons from the target atom by the projectile can equally well be said about the removal of electrons from the projectile by the target atoms. From the point of view of removing electrons from either the target or the projectile, what matters is that the two are in relative motion; which is actually moving in the laboratory frame of reference is unimportant. It follows that if the ionization potential of the projectile atom is I_p it will itself lose electrons in passing through the solid if its energy exceeds

$$E_s=I_pm/4m_e \tag{6.24}$$

The subscript s is used because this is a *stripping* process.

For a hydrogen atom I_p is about 14 eV, so that E_s is about 7 keV, which means that bare protons rather than hydrogen atoms will be the projectiles at all energies above 7 keV. In contrast, for argon E_s is about 300 keV, which means that even if *charged argon ions* are used

116

as projectiles, below 300 keV they will quickly collect electrons and pass through the solid as *neutral argon atoms*. In equation (6.14) we showed that the rate of energy loss to the electrons of a solid varied with the square of the projectile charge. It follows that little ionization will occur below E_s even if the projectile energy exceeds E_I, simply because its effective charge is zero.

What we have shown then is that below E_I and E_s heavy ions do not lose energy to electrons. Energy loss to electrons can occur at projectile energies above E_I and E_s, and by the time the energy is 2–3 times these values heavy ions behave just like light ones. Below the threshold energies essentially neutral atom–neutral atom collisions occur, so let us consider them in more detail.

Atom–atom collisions

For large enough separations two *neutral* atoms interact hardly at all. At smaller separations there is sometimes a covalent or hydrogen bonding force between them, and always the van der Waals force which varies with separation like R^{-7} (see Chapter 2). When the two atoms are very close together they always repel each other because the electron shells overlap.

At small separations ($R < R_o$, see figure 1.3), the repulsion always swamps the attractive forces and causes the atoms to behave as if they were rubber balls. The way the repulsive interaction energy changes with the separation of the atoms is not known at all well, but algebraic approximations to it are very useful.

One approximation is the screened coulomb potential, which derives from the assumption that the major repulsion comes from electrostatic forces between nuclei, and that the surrounding electrons merely become less able to screen the nuclei from one another as the separation is reduced. If no screening occurred at all, two atoms would have an interaction energy

$$U(R) = Z_1 Z_2 e^2 / \epsilon_o R \qquad (6.25)$$

and the repulsive force would be

$$F(R) = \frac{-dU(R)}{dR} = \frac{Z_1 Z_2 e^2}{\epsilon_o R^2} \qquad (6.26)$$

This force is, of course, the simple repulsion between a charge $Z_1 e$ and another $Z_2 e$. In the screened coulomb potential the interaction energy is reduced to

$$U(R) = \frac{Z_1 Z_2 e^2}{\epsilon_o R} \exp(-R/a_1) \qquad (6.27)$$

The screening parameter a_1 is always a small fraction of the size of the atom, so that when the separation is several times the atomic diameter

117

the interaction potential is much smaller than for an unscreened coulomb interaction. For small R the screening term is unimportant.

A second useful approximation to the repulsive interactions is the Born–Mayer formula:

$$U(R) = A_1 A_2 \exp\left(-R/a_2\right) \qquad (6.28)$$

where A_1 and A_2 are constants. In a way this interaction ignores the nuclear–nuclear repulsion but takes account of the repulsion between the electrons, which the screened coulomb potential ignores. As a result expression (6.27) is better at small separations, where the nuclear–nuclear repulsion dominates, but (6.28) is better when the atoms are farther apart. All three potentials, unscreened coulomb, screened coulomb and Born–Mayer for copper–copper interactions are shown in figure 6.13.

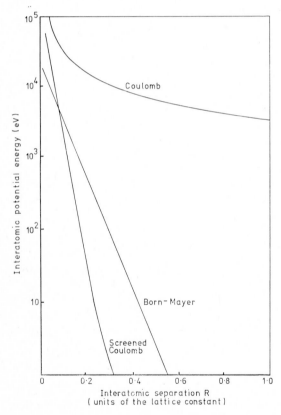

Fig. 6.13. Copper atom-copper atom interaction potentials (after G. Carter and J. S. Colligon, *Ion Bombardment of Solids,* Heinemann, 1968).

118

Because colliding neutral atoms behave like rubber balls, fairly violent collisions are common and a lot of energy can be lost by the projectile and gained by the target atom. What is more, atoms are fairly closely packed in solids so that collisions between the projectile and the atoms are very common, that is, the collision cross-sections are large, particularly if the particle energy is small so that it cannot approach the target atom very closely. If we look again at figure 6.13, and assume that the Born-Mayer potential is nearest reality, we will see that a projectile at 10 eV can approach another atom only to 0·4 of the normal lattice separation. The collision cross-section is therefore about $0·5\ a^2$, where a is the lattice constant. On the other hand, a projectile at 10 keV can approach as close as 0·05 of the lattice separation, and the collision cross-section is about $0·008\ a^2$. Since the cross-section decreases with increasing energy, high energy particles can pass much more easily into the solid. The ranges of heavy atoms are thus found to be roughly proportional to their energy. Nevertheless, even high energy heavy atoms undergo violent collisions and do not have very long ranges compared with electrons and protons. Table 6.1 shows the range of a variety of particles at 100 keV and illustrates this point.

Table 6.1.   The ranges of various particles at 100 keV in aluminium.

Electrons	40 μm
Protons	1 μm
Ar	60 nm
Kr	35 nm
Xe	28 nm

The generation of high energy particles

Natural sources of high energy particles are rather inconvenient. High voltage accelerators are far more powerful tools, since they provide high current, mono-energetic, mono-directional beams of almost any required particle and can be turned on or off at will. Their operation is conceptually very simple, although they are very often expensive and complex in practice.

The starting point is always the ion source, where the required atoms are converted into ions, usually by passing an electrical discharge through a suitable gas. If no suitable gas exists then the required ions can be made from a discharge in an inert gas (usually argon) by allowing inert gas ions to bombard a solid electrode and sputter atoms off it in the process. These sputtered atoms are then ionized in the inert gas discharge. If electrons rather than ions are needed a hot cathode is always used. Having produced the electrons or charged ions these are accelerated to the required energy in an electric field. This can be done in a single step or by making the particle move in a circle and giving it a large number of small accelerations which add together to give the required energy. Figure 6.14 shows a small 0·4 MeV particle accelerator used for radiation damage studies at Harwell.

119

Fig. 6.14. A 0·4 MeV particle accelerator at Harwell (courtesy of the U.K.A.E.A.).

6.4. *Neutrons*

Neutrons interact with solids quite differently from photons, charged particles or atoms. They have no charge, so that the electrical forces which have caused all the effects we have discussed so far in this chapter do not exist for them. They hardly notice the electrons of the solid at all, and react with the nuclei only through nuclear forces. These forces have a very short range, typically 10^{-14} m compared with the atomic size of $\sim 10^{-9}$ m, so that collisions between neutrons and nuclei are rare. Since the diameter of the nucleus is about 10^{-14} m (see eqn. (1.1)), the cross-section for neutron nuclear collisions is typically 10^{-28} m². These collisions are of crucial importance in nuclear reactors and a unit of cross-section called the 'barn' has become commonplace, with 1 barn $= 10^{-28}$ m². When collisions between neutrons and nuclei do occur they fall into one of two main types.

In simple collisions the neutron bounces off the nucleus just as a ball bearing might bounce off a billiards ball. The cross-section for these collisions is small, about 1 barn, but they are violent when they do occur and the neutron is likely to be deflected through a very large angle. As a result, the path of the neutron approximates to a random walk, that is the direction after a collision usually bears no relation to that before. The mean free path l is given from eqn. (6.9) by putting $x = l$ and setting $p = 1$. This means that l is given by

$$l = 1/N_0 \sigma \qquad (6.29)$$

so that for a typical solid where $N_0 = 4 \times 10^{28}$ m^{-3} and $\sigma = 10^{-28}$ m^2 the mean free path is 0·25 m. Stopping fast neutrons is indeed a formidable task.

Sometimes, however, the neutron is able to react with nuclei in the solid and cause a nuclear reaction. For example, slow neutrons, which have energies less than 1 eV, interact with ^{6}Li nuclei as follows:

$$^6\text{Li} + {}^1\text{n} \rightarrow {}^4\text{He} + {}^3\text{H} + 4·79 \text{ MeV}$$

The large amount of energy released is shared between the products of the reaction. Slow neutrons are much more effective at causing nuclear reactions than fast ones because they move past the nucleus slowly enough for reaction events to occur, whereas fast neutrons are more likely to interact only in simple collisions like billiards balls striking one another. Cross-sections for collisions between slow neutrons and nuclei can be very much larger than 1 barn. For example, that for capture of slow neutrons by ^{116}Cd is 25 000 barn. In this case the neutrons are actually absorbed by the ^{116}Cd so that we can go back to the concept of absorption coefficient that we used for photons and write

$$\alpha = N_0\sigma \ (\simeq 10^5 \text{ m}^{-1}) \tag{6.30}$$

This means that 1 mm of ^{116}Cd reduces the intensity of a beam of slow neutrons by a factor of $\exp(-100) \simeq 10^{-40}$, but will scarcely attenuate fast neutrons at all. This is the basis of a useful trick by which fast neutrons can be used for irradiation relatively free from slow neutrons.

Neutron sources

There are two important types of neutron source. In the first a high energy accelerator is used to generate particles which produce neutrons from a nuclear reaction when they hit a suitable target. For example, if ^{2}H$^+$ ions are accelerated to a few hundred keV and hit a target containing ^{3}H atoms then the following reaction occurs:

$$^3\text{H} + {}^2\text{H} \rightarrow {}^4\text{He} + {}^1\text{n} + 17·6 \text{ MeV}$$

The neutrons are created with an energy of about 14 MeV.

Neutrons in large numbers are best created in nuclear reactors. Here slow neutrons interact with heavy, fissile atoms such as ^{235}U or ^{239}Pu and cause them to break up. Usually each fission event creates two high energy heavy particles, called fission fragments, and several neutrons. The neutron energies are usually around a few MeV. A 20 MW materials testing reactor, such as those shown in figure 6.15 will generate such neutrons at a rate of about 10^{18} per second.

6.5. Summary

At this point it is worth summarizing the conclusions that we have reached about the way high energy radiation interacts with matter.

Fig. 6.15. Materials testing reactors at Harwell. The two reactors, Dido and Pluto, operate at powers of 20–25 MW and are used to test radiation effects in fuels and structural components. The reactor cores are only a few metres in diameter and length, but the grey tanks in which they are housed are about 15 m high (courtesy of the U.K.A.E.A.).

(i) Photons in the 10–100 keV range cause mainly photo-electric emission. In this process electrons are excited from *tightly* bound electron shells to highly excited levels, leaving energetic holes. The cross-section is highest for heavy atoms, varying with Z^4, and for low energies, varying with E^{-3}.

(ii) Photons in the range 100 keV to 1 MeV cause mainly Compton scattering. Here the incident photon is replaced by a scattered one of lower energy and the energy difference is given to an electron in the solid, usually from a *weakly* bound electron level. The cross-section per atom varies with $(Z \ln E)/E$ at sufficiently high energies (>1 MeV), falling much less dramatically with energy than photo-electric absorption.

(iii) Pair-production cannot occur below 1·022 MeV but quickly dominates photon–matter interactions above this energy. Electron positron pairs are created in the electrostatic field of the nucleus. The cross-section varies with Z^2 and rises rapidly with E above 1·022 MeV.

(iv) Light charged particles, such as electrons, protons and alpha particles, lose energy primarily to the electrons of a solid and in small packets, so that only slightly energetic electrons and holes are produced. Only a very small fraction of the energy is lost directly in collisions with nuclei in the solid. The range of these particles is roughly proportional to the square of their energy and inversely to the product of the electron density in the solid, the square of the particle charge and its mass.

(v) Heavy particles also lose energy primarily to electrons in the solid, provided their energy is above critical values E_I and E_s. Below E_I the particle is very unlikely to lose energy to electrons, and below E_s it is unable to remain electrically charged. As a result, heavy particles with energies lower than E_I and E_s move as neutral atoms and collide with atoms rather like rubber balls. Their energy is lost very rapidly in this way and their ranges are short.

(vi) Neutrons do not interact with electrons at all. They can collide with nuclei very much as billiard balls collide, but the cross-section is small and the range of neutrons consequently very long. Slow neutrons can also interact with nuclei and cause nuclear reactions to occur. Collision cross-sections can then be much larger.

In a sense this chapter has looked at the way high energy radiation interacts with matter from the point of view of the radiation. We have asked about the mechanisms of energy loss and the rate of energy loss, but have no more than touched on the fate of all this transferred energy. Often it causes radiation damage, and that we will discuss in the next chapter.

CHAPTER 7

the creation of defects by radiation

7.1. *Atomic displacements by direct momentum transfer*

It is intuitively rather obvious that when a high energy particle such as an electron or neutron strikes the nucleus of an atom, energy and momentum will be transferred, just as in a collision between billiards balls. Collisions of this kind are central in radiation-damage processes, and in discussing them we will always call the incident particle the *projectile,* and the atom with which it collides, the *target* atom.

If a projectile transfers only a small part of its kinetic energy to a target atom in a collision, then the atom will not leave its lattice site completely but merely vibrate around its equilibrium position a little more violently than usual. It is not displaced, for the simple reason that the energy transferred is insufficient to overcome the forces binding it to its lattice site. Eventually its excess energy is shared with the other atoms and appears as heat. On the other hand, collisions in which very large amounts of energy are transferred will almost certainly lead to displacement of the target atom. These statements lead us to the concept of a *displacement threshold energy* E_d; if the target receives energy less than E_d from the projectile it will not be displaced, but if it receives more than E_d it will be displaced. As a first guess we would expect E_d to be similar in size to the binding energy of the atoms in the solid.

In a real irradiation situation collisions between target atoms and projectiles take place in a variety of directions in the crystal. This complicates our concept of a displacement threshold energy because the energy required to displace the target atom will depend on the direction in which it first moves. Figure 7.1 shows this effect for a two-dimensional

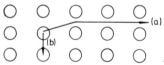

Fig. 7.1. Possible ejection paths for a target atom in a two-dimensional square lattice.

square lattice. If the target moves off along path (*a*) it can move relatively easily along the channel between two rows of atoms, whereas path (*b*) takes it directly into a head-on collision with its nearest neighbour. The displacement threshold for direction (*a*) should therefore be smaller than for direction (*b*), and we might expect to be able to define a curve such as that shown in figure 7.2 to represent the probability that a target atom which receives energy E will be displaced, assuming that the directions of collisions are random. However, this degree of refinement is not really necessary for our discussion since we want to outline principles rather than try to include all the details, so we will assume a single displacement threshold energy. We are, in effect, using the probability curve shown dotted in figure 7.2, and we shall find it a very useful simplification.

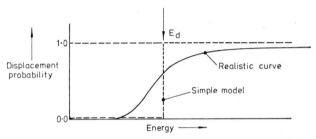

Fig. 7.2. The probability of displacement of a target atom after transfer to it of energy E.

Given the concept of a displacement threshold it is clearly very important to know whether the energy transferred to a target atom in a collision is smaller or larger than E_d, since only in the latter case is the taget atom displaced. It is worth considering this in two stages. First we need to discuss the *maximum* energy which the projectile can give to the target. We will show later that this is in a head-on collision; in glancing collisions much smaller amounts of energy are transferred. Secondly, we shall consider the probability of collisions in which a certain amount of energy is transferred and hence derive the average energy transfer.

Maximum energy transfer from projectile to target

It proves to be very easy to calculate the energy which is transferred in a head-on collision, using the laws of conservation of energy and momentum. Suppose, as in figure 7.3, we have a projectile of mass m_1, velocity v_o and kinetic energy $E_o(=\frac{1}{2}m_1 v_o^2)$. It collides head-on with a target atom having mass m_2 which is initially at rest, and after the collision the two particles have velocities v_1 and v_2 respectively. The law of momentum conservation tells us that

$$m_1 v_o = m_1 v_1 + m_2 v_2$$

and that for energy conservation requires that

$$\tfrac{1}{2}m_1 v_o^2 = \tfrac{1}{2}m_1 v_1^2 + \tfrac{1}{2}m_2 v_2^2$$

125

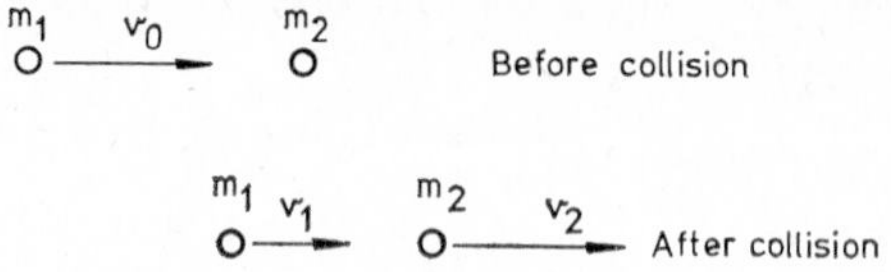

Fig. 7.3. Energy transfer in a head-on collision.

If we eliminate v_1 from these two equations we find that

$$v_2 = 2v_0 m_1 / (m_1 + m_2)$$

so that the energy transferred to the target atom is

$$E_2 = \tfrac{1}{2}m_2 v_2^2 = \tfrac{1}{2}m_1 v_0^2 \; 4m_1 m_2 / (m_1 + m_2)^2$$

Thus the maximum energy which can be transferred, which we will now call E_m, is related to the projectile energy by

$$\frac{E_m}{E_0} = 4m_1 m_2 / (m_1 + m_2)^2 \qquad (7.1)$$

In this derivation we assumed that all the particles were non-relativistic, that is moving slowly compared with the velocity of light. This is a very good approximation for all target atoms and for all projectiles except the electron. If the projectile is a non-relativistic electron $m_1 = m_e$ and we can assume $m_2 \gg m_e$. Equation (7.1) reduces to $E_m / E_0 \simeq 4m_e / m_2$. The relativistic equivalent is

$$\frac{E_m}{E_0} \sim \frac{4m_e}{m_2}\left(1 + \frac{E_0}{2m_e c^2}\right) \qquad (7.2)$$

Table 7.1 shows the maximum energy which can be transferred to hydrogen and copper atoms by three different projectiles of energy 1 MeV. It is clear from the table and from eqn. (7.1) that if the projectile and target have the same mass, then a head-on collision transfers the whole of the energy. On the other hand, if the projectile is light and the target heavy, then even a head-on collision does not transfer much energy; in table 7.1 we see that a 1 MeV electron can transfer only 69 eV to a copper atom.

Table 7.1. The maximum energies which can be transferred in a direct collision event.

Target atom	Projectile	1 MeV electron	1 MeV neutron	1 MeV copper atom
H atom		4·3 keV	1 MeV	61 keV
Cu atom		69 eV	61 keV	1 MeV

126

Measurement of displacement energy thresholds

The small maximum energy transfer from electrons to target atoms provides a convenient way of measuring displacement threshold energies. Threshold energies are usually around 20–50 eV and this order of energy can be transferred to atoms in head-on collisions with electrons having a kinetic energy of a few hundred keV. In the experiments the material being studied is irradiated with electrons and the incident electron energy increased until displacement of atoms (that is the creation of defects) can just be detected by some suitable means such as electrical resistivity (see Chapter 4). Suppose the critical electron energy at which displacement begins is E_c. This is related to the threshold energy E_d by eqn. (7.2), that is

$$E_d \simeq E_c \frac{4m_e}{m_2}\left(1+\frac{E_c}{2m_ec^2}\right)$$

In principle the same kind of measurement could be made using protons or other ions instead of electrons. In practice this is impossible because the critical proton energies required are a few hundred eV rather than the few hundred keV for electrons, and such low energy protons are much more difficult to handle and have an impossibly short range in solids.

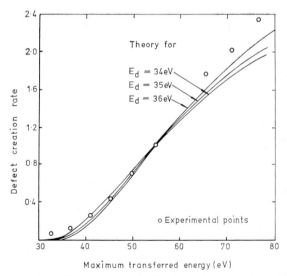

Fig. 7.4. Energy dependence of atomic displacement rates in gold (after W. Bauer and A. Sosin, *Physical Review*, **135**, A521 (1964)).

Figure 7.4 shows how the probability of producing displacements in gold by electron irradiation varies with E_m; remember that E_m is the

127

maximum transferred energy and is related to the incident projectile energy by eqn. (7.2). In this case the displacement rate was measured through changes in electrical resistivity, and the threshold energy indicated is about 30 eV.

Energy transfer in glancing collisions

If the collision between projectile and target is not head-on the transferred energy is less and both particles move off in directions away from the path of the incident particle, see figure 7.5. In this case it is a little less easy to determine the transferred energy because we do not immediately know the scattering angles η and ϕ in figure 7.5 (a). Nevertheless, the problem is not difficult if we consider the motion of the two particles relative to the motion of their centre of mass.

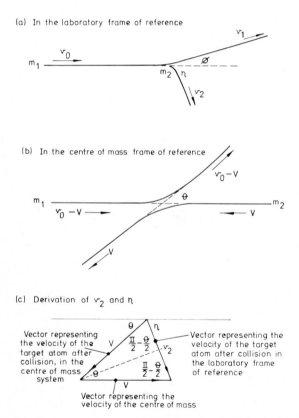

Fig. 7.5. The dynamics of a glancing collision in laboratory and centre-of-mass coordinates.

128

As before, we assume that the target atom is initially at rest and has mass m_2, while the projectile has mass m_1 and velocity v_0 before collision. The centre of mass of the two particles therefore has a velocity

$$V = v_0 m_1 / (m_1 + m_2)$$

in exactly the same direction as v_0. The kinetic energy of the two particles which is associated with the motion of their centre of mass is just

$$E_{com} = \tfrac{1}{2}(m_1 + m_2)V^2$$
$$= E_0\, m_1 / (m_1 + m_2)$$

and the remainder of the kinetic energy is associated with motion of the particles relative to their centre of mass. In the centre of mass frame of reference the projectile approaches the collision point with a velocity $v_0 - V$ and the target approaches with a velocity V from the opposite direction (see figure 7.5 (b)).

Because momentum and energy are conserved relative to the centre of mass the two particles must leave the collision with the same velocities as they approach it. This can be argued fairly simply as follows. Since the total momentum of the two particles is zero in the centre of mass system they must leave the collision in exactly opposite directions, and the ratio of their velocities must be the same after the collision as before. However both velocities cannot increase or decrease, because this would violate the conservation of energy. The situation is therefore as shown in figure 7.5 (b), with the particles being scattered through an angle θ relative to the centre of mass. It follows that v_2, V, θ and η are related as shown in the vector diagram of figure 7.5 (c), that is

$$\eta = \frac{\pi}{2} - \frac{\theta}{2}$$

and

$$v_2 = 2V \sin (\theta / 2)$$

The energy transferred to the target atom in the glancing collision is therefore

$$E_2 = \tfrac{1}{2}m_2 v_2^2 = \tfrac{1}{2}m_2\, 4 \left(\frac{v_0 m_1}{m_1 + m_2} \right)^2 \sin^2 \frac{\theta}{2}$$

$$= E_m \sin^2 \frac{\theta}{2} \tag{7.3}$$

As before E_m is the maximum energy which can be transferred, in a head-on collision with $\theta = \pi$.

In order to find the probability that a given amount of energy will be transferred to the target in a collision we need to know the probability that a given scattering angle will occur. If the probability of collisions which transfer an energy between E and $E+dE$ is $K(E)\,dE$ and that for collisions which result in scattering angles between θ and $\theta+d\theta$ in the centre of mass frame of reference is $\sigma(\theta)\,2\pi\sin\theta\,d\theta$, where E and θ are related by eqn. (7.3), then

$$K(E)\,dE=\sigma(\theta)\,2\pi\sin\theta\,d\theta$$

In this expression $K(E)$ and $\sigma(\theta)$ are the differential cross sections for transferring an energy E and for scattering through an angle θ, and we have included the term $2\pi\sin\theta$ because $\sigma(\theta)$ is usually given in terms of the cross-section per unit solid angle (steradian). If we use eqn. (7.3) to give $dE/d\theta=\frac{1}{2}E_m\sin\theta$ we have

$$K(E)=\frac{4\pi}{E_m}\,\sigma(\theta) \tag{7.4}$$

With this equation we can answer most of the questions about energy transfer, provided we know the cross-section $\sigma(\theta)$.

(a) *Hard-sphere scattering*

If the particles behave like hard spheres, then $\sigma(\theta)$ turns out to be constant. That is, any scattering angle is as likely as any other. As we saw in Chapter 6, there are two important examples of this kind of interaction in radiation damage; fast neutrons colliding with atomic nuclei and slow moving heavy ions colliding with other atoms. In the first case the magnitude of $\sigma(\theta)$ is about 1 barn per steradian and in the second case about 10^6–10^8 barns per steradian.

For hard-sphere scattering the total cross-section for displacement of the target atom is

$$\sigma=\int_{E_d}^{E_m} K(E)\,dE=\int_{E_d}^{E_m} 4\pi\,\sigma(\theta)\frac{dE}{E_m}$$

$$=4\pi\,\sigma(\theta)\,[1-E_d/E_m]\ \text{barn} \tag{7.5}$$

For collisions between neutrons and atomic nuclei in a solid the maximum transferred energy (E_m) is usually much larger than the displacement threshold energy (E_d) so that the total cross-section for displacement is about 10 barns. This means that a neutron incident on a solid with atomic density $\sim 4\times 10^{28}\ \text{m}^{-3}$ about 1 mm thick has a 4 per cent chance of displacing an atom. This follows from eqn. (6.9), which gives the probability of a particular collision as $N_o\sigma x=4\times 10^{28}\times 10^{-27}\times 10^{-3}=4\times 10^{-2}$.

The average energy transferred to those target atoms which are displaced in the collision is just

$$\bar{E}=\int_{E_d}^{E_m} K(E)E\,dE\ \Big/\ \int_{E_d}^{E_m} K(E)\,dE$$

$$=\tfrac{1}{2}(E_m+E_d) \tag{7.6}$$

130

We have already seen that E_m is large in neutron collisions (61 keV to Cu by a 1 MeV neutron) so that the average energy transferred is also large; about 30 keV in this case.

(b) *Rutherford scattering*

If the projectile is electrically charged, then the dominant force between it and the nucleus of the target atom is the coulomb force. This kind of interaction force dominates in irradiation by electrons and protons in particular, and its analysis was first carried out by Rutherford. Remember that we are trying to determine the probability of a given energy transfer $K(E)$ and we therefore need the probability of a given scattering angle θ in centre of mass coordinates. Rutherford showed that

$$\sigma(\theta) = \frac{s^2}{16}\left(\sin\frac{\theta}{2}\right)^{-4}$$

where the quantity s is the distance of closest approach of projectile and target and is given by

$$s = Z_1 Z_2 e^2 / 4\pi\epsilon_0 \cdot \tfrac{1}{2}M_r v_0^2$$

Here $Z_1 e$ = charge of projectile,

$Z_2 e$ = charge of target,

M_r = reduced mass of particles $(m_1 m_2/(m_1+m_2))$,

$\tfrac{1}{2}M_r v_0^2$ = initial kinetic energy in the centre of mass reference frame.

The Rutherford expression for $\sigma(\theta)$ is not difficult to derive, but the derivation can be found in many more advanced texts such as Goldstein's *Classical Mechanics* and is too lengthy to reproduce here. For Rutherford scattering, therefore, the total cross-section for displacement, remembering that $E/E_m = \sin^2(\theta/2)$, is

$$\int_{E_d}^{E_m} K(E)\,\mathrm{d}E = \int_{E_d}^{E_m} \frac{4\pi}{E_m} \frac{s^2}{16} \frac{E_m^2}{E^2}\mathrm{d}E$$

$$= \pi\frac{s^2}{4}\left(\frac{E_m}{E_d}-1\right) \tag{7.7}$$

For 1 MeV protons incident on copper $E_m = 61$ keV and $E_d \simeq 30$ eV. Since $s^2 \simeq 20$ barns the total cross-section in this case is 3×10^4 barns, which means that a 1 MeV proton is much more likely to displace a copper atom as it passes through a *thin* foil of the metal than is a neutron, where the total cross-section was only 10 barns. It does not follow that a 1 MeV proton will displace more atoms than a 1 MeV neutron in a thick solid, because of the much shorter range of the proton.

131

The average energy transferred in a Rutherford collision is

$$\bar{E} = \int_{E_d}^{E_m} K(E)E \, \mathrm{d}E \bigg/ \int_{E_d}^{E_m} K(E) \, \mathrm{d}E$$

$$= E_d \ln (E_m/E_d)/[1 - E_d/E_m] \tag{7.8}$$

Using again the example of 1 MeV proton irradiation of copper the average energy given to a displaced atom is only 230 eV, much smaller than for hard-sphere collisions.

Numbers of primary displacements

We are now in a position to calculate the number of primary displacements caused by a particular kind of particle irradiation, and we can illustrate the calculations with the examples of neutron and proton irradiation.

(a) Reactor neutron irradiation

In a nuclear reactor there is always a fairly high flux of fairly energetic ($\sim$1–5 MeV) neutrons. It is usual to express this flux in terms of the number (ϕ) of neutrons crossing unit area of any given plane in unit time, and to assume they all have a mean energy E_0. If we then consider a section of material with thickness x, unit area and atomic density N_0, and the cross-section for displacement per incident neutron is σ, we can write the total number of displacements in time t by using eqn. (6.9). If we also substitute eqn. (7.5) for σ we have

$$n = N_0 x \, 4\pi \, \sigma(\theta)[1 - E_d/E_m] \, \phi t$$

Thus the probability that any given atom will be displaced by a neutron in a primary collision is just

$$n/N_0 x = 4\pi \, \sigma(\theta) \, [1 - E_d/E_m] \, \phi t \tag{7.9}$$

A typical materials testing reactor has $\phi \sim 10^{16} \, \mathrm{m}^{-2} \, \mathrm{s}^{-1}$, so that a 10-day ($10^6$ s) irradiation displaces about one atom in every 10^5 by *primary* collisions, assuming $\sigma(\theta) \sim 1$ barn.

(b) Proton irradiation

Protons that are used to create radiation damage in solids usually have energies in the range 0·1 to 5·0 MeV, and ranges in solids from 2 to 200 μm. For this reason the solids used are almost always thick enough to stop the protons completely, and in calculating the number of primary displacements per incident particle we must allow for the loss of energy as the proton approaches the end of its range. Provided the way the proton loses energy with distance is known the number of displacements can be calculated exactly.

132

If the proton loses energy uniformly as it penetrates the solid, the number of primary displacements is

$$N_o R \frac{\pi}{4} s_o^2 \frac{E_m}{E_d} \ln \frac{E_m}{E_d} \tag{7.10}$$

N_o is the density of atoms as before, R is the range of the proton, E_m is the maximum energy transferred to the target by a proton of energy E_o, and s_o is the value of s for a proton of this energy. For a fairly heavy target atom $M_r \simeq m_1$ and

$$s_o \simeq Z_2 e^2 / 4\pi\epsilon_o E_o \tag{7.11}$$

Again using the example of 1 MeV protons incident on copper we find that the total number of primary displacements per incident proton is about 10 since

$$N_o \simeq 8 \times 10^{28} \text{ m}^{-3}$$

$$R \simeq 5 \ \mu\text{m}$$

$$s_o^2 \simeq 20 \text{ barn} = 2 \times 10^{-27} \text{ m}^2$$

$$E_m \simeq 61 \text{ keV}$$

$$E_d \simeq 30 \text{ eV}$$

The equivalent calculation for electron irradiation is very similar but complicated by having to allow for the relativistic effects of eqn. (7.2).

Displacement cascades

We have now looked in some detail at the way energy is transferred from the projectile to the target atom in a direct collision event. We have seen how we can calculate the number of atoms displaced by these collisions. But we have not yet taken account of the fact that the atom displaced in the direct collision event may itself have enough energy to displace other atoms in secondary collisions. These atoms may in turn replace yet more atoms. This chain of events is called a *collision cascade* and is illustrated in figure 7.6.

Suppose the primary collision transfers enough energy to the target atom for it to leave its site with energy E, and that it creates $n(E)$ displacements in the subsequent collision cascade. In its first collision with another atom it is scattered with an energy E_1 and the second atom is given an energy $E-E_1$. If the second atom leaves its lattice site, that is it is displaced, it will retain an energy of about $E-E_1-E_d$ and one defect will have been created. The collision between the first and second atoms will usually be hard sphere in character because, as we saw in Chapter 6, heavy atoms collide in this way unless their energies are very high. For hard-sphere collisions $\sigma(\theta)$ is constant and it follows

133

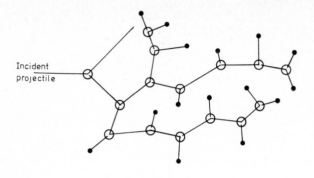

O = Vacancy defects
• = Interstitial defects
——— = Ion tracks

Fig. 7.6. A displacement collision cascade.

from eqn. (7.4) that any given energy transfer in a collision is as likely as any other. Since the primary target atom, which initially had an energy E, can lose any energy between E and 0 to another similar atom (with the same mass) it follows that the probability that it will lose an energy between $E-E_1$ and $E-E_1+dE_1$ is just dE_1/E. It then follows that

$$n(E)= \int_0^E n(E_1)\frac{dE_1}{E} + \int_0^E n(E-E_1-E_d)\frac{dE_1}{E} +1 \qquad (7.12)$$

This equation says that the number of displacements $n(E)$ produced by a primary target atom which is given energy E is equal to the sum of the probable number of subsequent displacements by both primary and secondary target atoms, averaged over all possible ways of sharing the energy, plus 1. The extra 1 takes account of the fact that the displacement of the secondary target atom has to be included. If we write $E/E_d=X$ and multiply by X then eqn (7.12) becomes

$$X n(X)= \int_0^X n(X_1)\, dX_1 + \int_{-1}^{X-1} n(Y)\, dY +X \qquad (7.13)$$

where $Y=X-X_1-1$. Differentiating this equation with respect to X gives

$$X\frac{dn(X)}{dX}+n(X)=n(X)+n(X-1)+1 \qquad (7.14)$$

For large n we can make the approximation that $n(X-1)+1\simeq n(X)$. This is a good approximation because, when X and $n(X)$ are large, the fraction of all collisions which transfer less energy than E_d is small,

134

namely $(1-1/X)$ for hard-sphere collisions. We then have the differential equation

$$X\frac{\mathrm{d}n(X)}{\mathrm{d}X}=n(X) \qquad (7.15)$$

for large X, and this has the solution

$$n(X)=\alpha X, \text{ where } \alpha \text{ is a constant}$$

Putting in appropriate boundary conditions for small X and solving the differential equation properly shows that the constant α is about 0·5, and gives a variation of $n(X)$ with X as shown in figure 7.7.

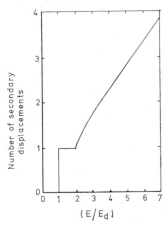

Fig. 7.7. The number of displacements produced by a primary of energy E.

In our model, in which an energy E_d is required for creating one defect, the absolute maximum number of additional defects which could be created by a displaced atom with energy E is $E/E_d\ (=X)$. We have just shown that only about half that number are actually created, the remaining energy being transferred to atoms in the solid in amounts smaller than E_d and therefore not creating any defects. This displacement cascade contribution to defect creation is extremely important in neutron irradiation, less so in proton irradiation, and can safely be ignored in electron irradiation. For example, in table 7.1 we noted that a 1 MeV neutron could transfer up to 61 keV to atoms in copper. If we again assume $E_d\sim30$ eV the average energy transferred in a primary collision (eqn. (7.6)) is about 30 keV. Thus for the average colision $X\sim1000$ and we would expect 500 secondary defects to be created. For 1 MeV protons E_m is still 61 keV, but the average energy transferred is only 230 eV (eqn. (7.8)) which means that, on average, only four secondary defects are created per primary displacement. For electrons at

135

1 MeV incident on copper the maximum energy transfer is only 69 eV, and the average energy transfer little more than E_d, so that secondary displacements are unlikely.

Practical relevance

Our society is an increasingly voracious consumer of energy, and in the future more and more of it will be generated in nuclear or thermo-nuclear reactors. The intense neutron and gamma-ray fluxes inside these reactors inevitably cause serious radiation damage in the fuels, cladding materials and other parts of the structure, and always the mechanism of damage is just the direct collision events we have analysed in this chapter. For this reason enormous efforts have been expended in trying to understand radiation damage and the properties of irradiated materials, often using aspects of the analysis we have just carried out. We will look at this in much more detail in Chapter 8, but it is perhaps worth pointing out here how the analyses we have just sketched out bear on research for the nuclear industry. For example, electrons are always used to generate the simplest irradiation products, single vacancies and interstitials. They do this because the maximum and average energies transferred in primary collisions are small so that displacement cascades rarely occur at all and are always very small. With electron irradition a detailed picture of the properties of defects has been built up which has proved invaluable in predicting the behaviour of materials in reactors.

Another example is the use of heavy ion irradiation to simulate the effects of very long reactor neutron irradiations. During the 20–30 year design lifetime of a high power reactor every atom in some parts of the structure will be displaced several times, and the engineers who must know the behaviour of the materials under these extreme conditions clearly cannot afford to wait 20–30 years for design information. Heavy ion irradiation has provided a useful solution. From our discussion of collision cascades we might expect a 30 MeV copper ion incident on copper to produce $\sim \frac{1}{2} \times 30 \times 10^6 / 30 = 5 \times 10^5$ displacements. Since the range of 30 MeV copper ions is about 3 μm it follows that a current of 10 mA m^{-2} of 30 MeV copper ions will create defects at the rate of 10^{28} m^{-3} s^{-1}, so that every atom is displaced in about 10 s. This calculation is crude and too optimistic, but it does serve to show that a few minutes' irradiation with heavy ions from an accelerator can simulate tens of years in a power reactor.

The radiation-damage events we have described in Section 7.1 are conceptually very simple. The irradiating particles collide with the atoms or nuclei in the solid and, if enough energy is transferred in the collision, the target atom is displaced. We have not had to consider at all the electrons in the target solid, except in so far as it is the electrons which bind the atoms and therefore determine E_d. In the next section we will look in detail at what happens to the electrons in irradiated solids and how their rearrangement can often contrive to move atoms and create atomic defects. We saw in Chapter 6 that all particles except neutrons and slow-moving heavy ions lose most of their energy to the electrons of solids; it follows that effects arising from electron excitation can sometimes dominate radiation damage.

7.2. *Electronic rearrangements*

High energy photons and charged particles interact primarily with electrons in solids; as we saw in Chapter 6 they usually excite electrons from one of the filled bands of the solid into the unbound continuum of levels above the conduction band. What follows on after this depends on the kind of material we are irradiating, for example whether it is a metal or an insulator, and what kind of excitation event has taken place. It proves to be useful to subdivide the relaxation processes which follow an electronic excitation event into two classes, those which occur so quickly that they can be thought of as instantaneous and those which may on occasion occur very slowly and even be permanent under suitable conditions.

Fast electronic relaxation processes

It is probably worth reminding ourselves at this stage of the character of electron energy bands in solids, which were introduced in Chapter 2 and used again in Chapter 6. In most solids there are a number of filled electron bands with a variety of binding energies. The most tightly bound electrons are in bands which correspond to the innermost electron shells (the K shells) of the constituent atoms, and they rarely contribute significantly to the binding energy of the solid. It is the electrons which are less tightly bound to the individual atoms and form the valence and conduction bands which usually dominate the binding energy of the crystal. In insulators and semiconductors the valence band is completely filled and the conduction band completely empty. In metals, on the other hand, the conduction band is partly filled. High energy states in the conduction band correspond to larger and larger electron kinetic energies; eventually the kinetic energy can be large enough for the electron to leave the crystal altogether if it happens to be near a surface. The energy level at which this becomes possible is called the vacuum level because here the electrons have the same energy as a stationary electron in a vacuum. If, however, an excited electron is well within a bulk metal, then energies even higher than the vacuum level can be achieved, in fact there is an infinite continuum of possible states extending upwards in energy from the vacuum level.

In a general electronic excitation process caused by a high energy photon or charged particle an electron will be excited from one of the deep, filled bands, such as that corresponding to the K atomic electron shells, into the continuum of unbound energy levels above that of the vacuum. The first thing that happens then is that the energetic electron and hole share their energy with the other electrons in the solid. The energetic electron may excite yet another electron from a filled band, perhaps the valence band, to the empty continuum. The energetic hole may be filled by an electron from a higher band and the energy released as an X-ray photon or used to cause further electron excitation as illustrated in figure 7.8. In this way a single high energy electronic excitation is soon degraded into a large number of electronic excitations each of much lower energy, and the time taken is typically 10^{-15} s.

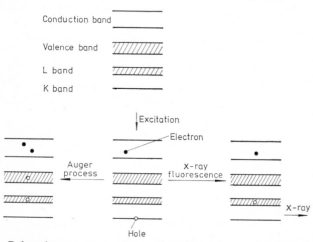

Fig. 7.8. Relaxation processes when a hole is created in a deep energy band.

In metals this process of energy sharing with other electrons can continue indefinitely, since even slightly excited electrons can lose part of their energy to certain electrons in the conduction band, where large numbers of electron states are empty just higher in energy than other occupied states. As a result an electronic excitation in metals is soon completely dissipated as heat. In insulators and semiconductors the sharing process cannot continue indefinitely because once an energetic electron has insufficient energy to excite another electron across the gap from the valence band to the conduction band it can no longer lose energy to electrons. After this point has been passed the electron loses energy more slowly, creating phonons as it does so, but remains in the conduction band. Similarly, holes remain in the valence band. The net result here is that a single excitation event of energy E_o is quickly transformed into a large number of holes in the valence band and electrons in the conduction band. If the band gap is E_g, then about

$\frac{1}{3} E_o/E_g$ electron–pairs are usually created, the remainder of the energy being lost in phonon creation.

It is therefore not surprising that electronic radiation effects do not occur in metals, and we need not consider them further. In insulators and semiconductors, on the other hand, electronic rearrangement can be very important.

Electron and hole trapping

In most insulators and semiconductors both electrons and holes are mobile at room temperature. Since they are attracted to each other by the coulomb force between them they eventually recombine. Very often the recombination energy is released as a photon; for example, light emission is a primary result of electron–hole recombination in zinc sulphide which is used as a phosphor in TV-picture tubes. Sometimes, however, the recombination energy creates only phonons in which case we have a non-radiative recombination event.

If the solid contains impurities which can trap both electrons and holes, they can be kept far enough apart to prevent their recombination occurring at all. One simple example of this situation occurs in KI containing a little TlI and held at a low temperature (between 150 K and 200 K). Thallium iodide dissolves in KI in such a way that some of the K^+ ions are replaced by Tl^+. The substitutional Tl^+ ions can trap electrons, becoming Tl atoms, and they can also trap holes to become Tl^{2+}.

The result is that a KI crystal containing TlI, when irradiated at about 150–200 K, is found to contain both Tl atoms and Tl^{2+} ions. The presence of Tl atoms can readily be detected through their optical absorption spectra, as shown in figure 7.9 curve (b). In both unirradiated and irradiated crystals there is a strong absorption band due to Tl^+, but only in the irradiated crystal can the band due to Tl atoms be seen. If the irradiated crystal is heated to room temperature the Tl atoms lose electrons and the Tl^{2+} ions lose holes, the electrons and holes recombine and the crystal returns to the virgin state. Recombination can also be achieved with the aid of photons which can be absorbed by Tl atoms, in the $Tl^°$ band indicated on figure 7.9, and ionize them. The band labelled F in figure 7.9 is due to F centres also formed by the X-irradiation. Their formation will be described later.

In some insulators, such as alkali halides at low enough temperatures, it is not necessary to have impurities in order to trap holes. In these materials the hole in the valence band causes the ions near it to move away from their normal lattice positions, as shown in figure 7.10. In effect the hole combines with two halide ions to give a halogen molecule ion (such as Cl_2^- in NaCl) in which the halogen nuclei are closer together than in the perfect lattice. The resulting 'self-trapped' hole cannot move

139

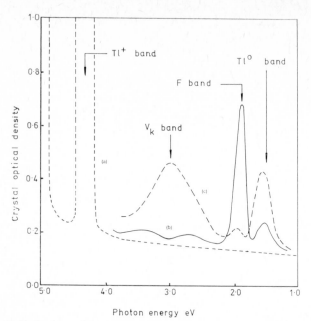

Fig. 7.9. Optical absorption spectra of KI containing about 0·1 per cent TlI. Curve (*a*): before irradiation; curve (*b*): after irradiation at 150-200 K; curve (*c*): after irradiation near 4 K.

very easily even at high temperatures, because heavy ions have to be moved as well as light electrons, and at low temperatures it does not move at all. Figure 7.9 curve (c) shows what happens to the optical absorption spectrum of KI containing 0·1 per cent Tl when irradiated near 4 K, the temperature of liquid helium. The electrons are trapped at Tl$^+$ ions as before, to give the Tl atom band. But now the holes are self-trapped and give rise to the optical absorption band labelled V$_K$.

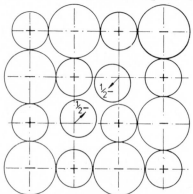

Fig. 7.10. The self-trapped hole in alkali halides.

140

Since electron and hole trapping effects simply involve moving electrons around in the solid they are usually not very permanent. The effects we described in KI:Tl are not stable at all at room temperature. However, although rarely permanent they are usually created very efficiently, because they occur after any electronic excitation event and it is in such events that ionizing radiation loses most of its energy. Because of their efficiency they are potentially useful in information storage devices and in radiation dosimetry, as we shall see in Chapter 9.

7.3. *Photochemical processes*

In most insulators such as oxides and diamond, and in semiconductors like silicon and germanium, electronic excitation can only lead to electronic rearrangement. In these materials ions are displaced and ionic defects formed only by the direct collision processes we talked about in Section 7.1. In some insulators, on the other hand, electronic excitation actually does cause ions and atoms to be displaced from their perfect crystal positions. The most important class of materials in which this kind of process occurs is organic compounds like sugar. Another very interesting class is alkali halides like common salt.

In the alkali halides the vacancy which is left behind when a halogen atom is moved in this way has an electron left trapped in it, and these vacancies plus electrons are the well-known F centres which give irradiated alkali halides their magnificent colours. In organic compounds hydrogen atoms are often ejected from the molecules by the radiation and very complex molecular fragments are left behind.

The nature of this kind of radiation-damage process is now fairly well understood but is rather complex. The basic principles involved can best be illustrated by looking at the photochemical dissociation of halogen molecules, and in order to do this we need to remind ourselves of the nature of electron energy levels in molecules. In molecules, just as in atoms, the electrons can occupy only well-defined energy levels, but in a molecule like Cl_2 the energy of these levels depends on the distance between the two atoms, as shown in figure 7.11. A diagram like figure 7.11, which plots the total energy of various electronic states of the molecule as a function of nuclear separation, is sometimes called a configuration coordinate diagram. The normal ground electronic state of the Cl_2 molecule is represented by the line (a) and has an equilibrium internuclear separation R_0. The binding energy of the molecule (E_B) is the energy required to move from the equilibrium point (A) to the dissociated state (point C).

If the molecule is excited from point A to a higher electronic state (point B) there will usually be a repulsive force between the atoms (the molecule is in an *antibonding* state) and the molecule will dissociate. This is because the molecule at point B has excess potential energy, so

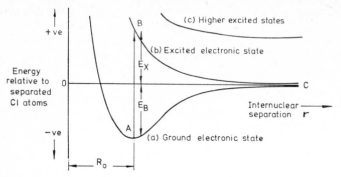

Fig. 7.11. The total energy of the Cl_2 molecule as a function of Cl–Cl distance for three electronic states.

that whereas in the ground electronic state (a) on figure 7.11 configuration A is adopted in the excited electron state (b) the molecule moves to configuration C. When the molecule is excited from A to B and dissociates each atom will have a kinetic energy equal to $\frac{1}{2}E_X$. This, then, is the essence of photochemical dissociation, and illustrates how the break up of a molecule can be caused by an electronic excitation, which did not initially affect the nuclei at all.

In alkali halides the photochemical damage process is somewhat similar to the break-up of chlorine molecules. The ionizing radiation first creates electron–hole pairs and the holes are quickly self-trapped as in figure 7.10. When electrons recombine with these self-trapped holes they can do so in a non-radiative process direct to the ground electronic state. When this occurs the two halide ions, which had previously made up the Cl_2^- molecule ion of the self-trapped hole, find themselves too close together and proceed to fly apart, just as the chlorine atoms fly apart during the break-up of the excited Cl_2 molecule. In a significant fraction of such events enough energy is released to throw one of the halogen ions permanently from its lattice site. As we might expect the real damage process is rather more complicated than this, but its fine details are not important to us here. The important points are that a single electron–hole pair can create an ionic defect, and that electronic potential energy can readily be converted into kinetic energy of the ions because of the way the total energy of molecules varies with nuclear configuration. As a result photochemical-damage processes can be very efficient, often nearly as efficient as electron rearrangements, and can usually be effected by ultraviolet irradiation with enough energy to create electron–hole pairs. In molecular solids such as organic compounds the situation is even more complex, because so many more atoms are closely involved, but the principle is the same.

142

Reasons for the rarity of photochemical damage

We have already said enough about photochemical damage processes to know why they never occur in metals. Going back to figure 7.11 we see that the lifetime of the excited state must be long enough for the two Cl atoms to separate, otherwise the molecule will move from curve (b) to curve (a) before full separation has occurred and then return to the bound configuration. In metals all excited electronic states with significant energy do have very short lifetimes; energy is so easily lost to the conduction electrons that there is never time for photochemical processes to occur.

In ceramic oxides like MgO photochemical processes do not occur. Again the reason is fairly clear, though quite different from that preventing photochemical damage in metals. Here the problem is essentially that electron–hole pairs do not have enough energy to do the job. We saw in Chapter 2 that the binding energy of ions in MgO was $\sim$30–40 eV because of the double charges involved. We also argued at the beginning of this chapter that the energy required for displacement would be about the same as the binding energy. Since electron–hole pairs in this material have energies of $\sim$7 eV it is scarcely surprising that photochemical damage cannot occur.

Photochemical damage is usually not important in semiconductors, in this case partly because the electron excitations are too widely dispersed. In alkali halides the self-trapped hole is confined to two ions, so that when recombination occurs these take the lion's share of the energy available and can therefore be displaced. In silicon and germanium electrons and holes move freely through the lattice and do not interact strongly with particular ions. In any case their energy ($\sim$1 eV) is in insufficient to create displacements.

7.4. Radiation damage by nuclear reactions

We have already thought about radiation damage by fast neutrons, when the neutron collides with a nucleus in the solid and transfers enough energy to cause a displacement. Neutrons can also react with the nuclei of target atoms and cause nuclear reactions to occur. Generally large quantities of energy are released by these reactions and given to their products. The reaction products can then cause direct collision displacements, electronic rearrangements or photochemical damage.

Slow neutrons, which cannot themselves cause damage, are often very effective at starting nuclear reactions. For example, in the reaction

$$^6\text{Li} + {}^1\text{n} \rightarrow {}^4\text{He} + {}^3\text{H} + 4 \cdot 79 \text{ MeV}$$

about three-sevenths of the energy released is given to the helium nucleus and four-sevenths to the tritium nucleus, which become highly damaging projectiles. The most important nuclear reactions from the point of view of radiation damage are those which involve fission of ^{235}U or ^{239}Pu, since these reactions are the basis of nuclear power generation. The products of these fission reactions vary, but the most likely ones are two nuclei with masses near 95 and 140, together with two or three neutrons. About 200 MeV is released in the reaction and shared among the products, which can clearly cause large numbers of defects to be formed (see also Chapter 8).

7.5. Secondary processes

So far in our survey of radiation damage we have thought a lot about how single ions or atoms are moved from their lattice sites, in primary collisions or in subsequent cascades, in photochemical processes or via nuclear reactions. We have said very little of what happens to them afterwards. This is an important omission because in many cases what happens afterwards is critical in determining the properties of the irradiated solid.

Athermal annealing

At sufficiently low temperatures both interstitial ions and vacancies are frozen in the crystal lattice, that is they cannot move because insufficient thermal energy is available for atomic jumps. It is therefore not very easy for interstitials to recombine with vacancies under these conditions, since neither can move, and we would expect to be able to build up very large concentrations of defects. This is what does sometimes happen! In metals at low temperatures electron irradiation can build up very much larger concentrations of defects than at higher temperatures. However, the concentrations are still limited, by a process called athermal annealing. This phrase is used to describe how an interstitial and a vacancy, if they are sufficiently close together, will recombine with each other even at the lowest temperatures. Athermal annealing occurs because interstitial atoms and vacancies attract one another through their respective strain fields and, when they are close enough, the attractive force is strong enough to overcome those forces holding the interstitial in a particular site. Figure 7.12 illustrates this by showing schematically the energy of an interstitial-vacancy pair as a function of the separation of the interstitial atom from the vacancy. Six sites which would normally be satisfactory for interstitial atoms are shown, and indeed sites 4, 5 and 6 are not very different from interstitial sites far away from any vacancy. That is, the atom is bound at these sites with an energy ΔE and cannot move at very low temperatures. Site 3 is not normal in that motion towards the vacancy can occur with an activation energy much smaller than normal, and sites 1 and 2 are not stable at all. If a newly created interstitial comes

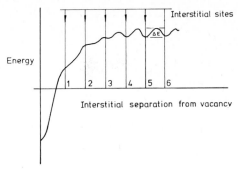

Fig. 7.12. The energy of an interstitial vacancy pair as a function of separation.

to rest in a site near enough to an already existing vacancy, in a site of type 1 or 2, then the two defects will annihilate each other and no new defect will have been created. It is fairly clear that athermal annealing will become very effective when defect concentrations are high, since then a new interstitial is almost certain to find itself close enough to an already existing vacancy for annihilation to occur. It is one of the important processes which limits defect concentrations at high irradiation doses.

Thermal annealing

If the temperature of irradiation is raised the defects actually become mobile. It is generally found that interstitials move more readily than vacancies, but if the temperature is high enough vacancies as well become mobile. If either or both defects can move, a considerable number of the defect pairs originally created may recombine and the apparent defect creation rate may be much less, even though the defect concentration is nowhere near large enough for athermal annealing to occur. Between the extreme situations of solely athermal annealing and full defect mobility there are many intermediate cases. If we return to figure 7.12 we can see that there will be a temperature lower than that at which full interstitial diffusion can occur but high enough to allow interstitials created in sites of type 3 to recombine with the nearby vacancies.

In copper there is a whole spectrum of sites like 3. In irradiations at the lowest temperatures a large number of defects are created, but if after irradiation the temperature is raised more and more pairs of interstitials and vacancies can recombine, so that the net defect concentration falls dramatically. This is illustrated in figure 7.13, and the behaviour of copper is not untypical. A curve similar to that shown in figure 7.13 could also be obtained by irradiating a crystal at various temperatures, and plotting the defect creation rate as a function of temperature.

145

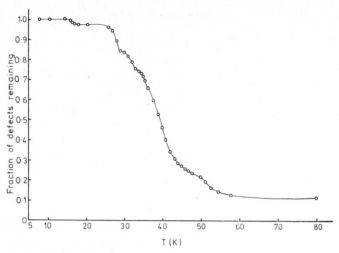

Fig. 7.13. The annealing of defects in copper after irradiation with 1·4 MeV electrons at about 8 K. The fraction of defects remaining after warming the sample of the temperature shown was measured by monitoring changes in electrical resistivity (after J. W. Corbett, R. B. Smith and R. M. Walker, *Physical Review*, **114**, 1452 (1959)).

The formation of clusters

So far we have considered only secondary processes in which defect pairs are annihilated. These are sometimes called back-reactions because they drive the irradiated crystal back towards the perfect state and act in opposition to the radiation-damage processes. Their effect is therefore solely to reduce the net defect creation rate from that which would have been expected from knowing the number of primary and secondary displacements. The reduction can be spectacular, as we have seen in the case of copper, but it will not, in itself, change the character of the radiation damage very much.

However, once defects become mobile, they can collect together into clusters as well as merely moving to annihilate one another. If copper is irradiated at room temperature, or even as low as 80 K, interstitials are mobile. Many interstitials find vacancies, and mutual annihilation occurs, but some find other interstitials and form interstitial dislocation loops. These are planar clusters of extra atoms as was shown in figure 3.4 and are very easy to see with an electron microscope. They form in ionic solids as well (figure 7.14). It is the formation of clusters which is most important in determining the change in the properties of reactor materials caused by radiation, and this aspect will be the subject of the next chapter.

It is perhaps worthwhile summarizing the major points which have been made about radiation damage in this chapter:

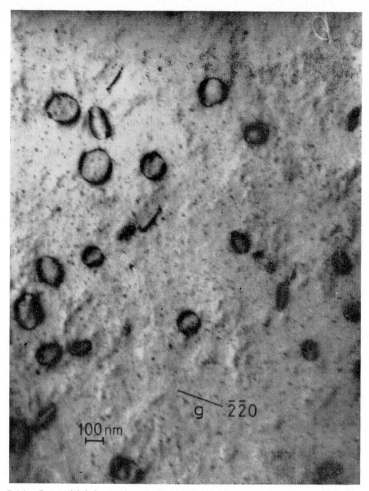

Fig. 7.14. Interstitial loops in alkali halides (after L. W. Hobbs, A. E. Hughes and D. Pooley, *Proceedings of the Royal Society*, **A332**, 167 (1973)).

(*a*) Atomic displacements in metals and many other solids can occur only by direct collisions between radiation particles and the atoms in the solid.

(*b*) Heavy projectiles are much more effective at producing these displacements than light ones. For light particles, particularly electrons and protons, the efficiency of direct displacement is low.

(*c*) Atomic displacement in some solids, such as alkali halides, can occur by a photochemical process, in which the radiation is required only to excite electrons as an initial step. If such a process does occur

147

the efficiency of defect creation can be very high, even under electron or proton bombardment.

(*d*) In insulators and semiconductors under suitable conditions, particularly when electron and hole trapping impurities are present, semi-permanent electronic rearrangement can occur under irradiation.

(*e*) The net rate of defect creation and the kind of defects formed are often critically dependent on the reactions between the simple defects which occur after their creation by the radiation.

CHAPTER 8

technological problems in radiation damage

Most of the impetus for studies of radiation damage in materials stems from the severe irradiation conditions present in nuclear reactors. It might be thought that since reactors have now been with us for a quarter of a century, most of the radiation damage problems associated with them would have been solved. However, despite the large number of scientists who have been engaged in this area, many basic aspects of radiation damage remain poorly understood, and the continual development of new reactor types throws up new and often unforeseen problems. In this chapter we attempt a brief review of some of the technical aspects of radiation damage in reactor environments, as well as some miscellaneous examples of the importance of radiation effects in other situations. It is impossible to do justice to this large topic in one chapter, and indeed a treatment which included very much scientific and engineering detail would be well outside the scope of this book. Our discussion will therefore be a general overview of the subject, beginning with some background material on reactor physics.

8.1. *Radiation in nuclear reactors*

Fission reactors

All nuclear reactors in operation today, and envisaged for the near future, are *fission reactors*. That is, they work by sustaining a chain reaction involving the fission of heavy nuclei such as ^{235}U. In a fission reaction a neutron penetrates the nucleus and causes it to break up into two lighter nuclides (fission fragments), at the same time releasing energy and some further neutrons. These neutrons are able to go on and produce fission in further nuclei, thus sustaining a chain reaction. The total energy released in a single fission event is about 200 MeV. In a nuclear reactor the rate of energy released by the fission process, which appears initially mostly as kinetic energy of the fission fragments, is controlled by varying the number of neutrons present in the system. This is done by 'control rods' which are made of neutron-absorbing material, and which may be inserted into the reactor as required. The kinetic energy of the particles released by fission is spent in producing radiation damage and heat in the materials of the reactor. The radiation damage constitutes a nuisance factor, but the heat can be used to raise steam for driving turbines to produce electricity or, as in a nuclear-powered ship, mechanical power. In an atomic bomb the energy released by fission is uncontrolled, and the result is a very powerful explosion.

The naturally occurring fuel for fission reactors is uranium. The most abundant isotope is ^{238}U, ^{235}U being present at the level of only 0·7 per cent and ^{234}U at 0·006 per cent. Fission of ^{238}U is only possible with neutrons of energy greater than about 1 MeV, whereas fission in ^{235}U is possible with neutrons of all energies. Although most of the neutrons released in the fission process have energies in the region of 1–2 MeV, they lose energy rapidly by collisions, and as a result it is not possible to sustain a chain reaction in the isotope ^{238}U. A further snag is that neutrons in the energy range 6–200 eV are captured by ^{238}U (leading to ^{239}U and ultimately, by radioactive decay, to ^{239}Pu) and become unavailable for producing fission in ^{235}U. A chain reaction can therefore only proceed in natural uranium if the neutrons are quickly slowed down to have energies below 6 eV, and then allowed to produce fission in the 0·7 per cent abundant isotope ^{235}U. This is accomplished by the use of a *moderator,* a material in which the neutrons may lose energy rapidly by collision. From eqn. (7.1) it can be seen that the energy lost by a neutron in a collision is greatest if the target atom has a low atomic mass, so the most efficient moderators are light materials. In practice water (H_2O), heavy water (D_2O) and graphite (C) are used, although H_2O has the disadvantage that hydrogen has a high neutron capture cross-section of 0·33 barn. It turns out that, because of this, reactors with H_2O moderators must use uranium enriched with the isotope ^{235}U in order to have an increased fission probability to overcome the effects of the absorption of neutrons by the H_2O. By using a moderator the neutrons may be slowed down until they have thermal energies, of $\sim kT$, where T is the temperature of the moderator. At reactor operating temperatures of a few hundred °C this energy is $\sim 0·05$ eV. These 'thermal neutrons' are able to cause fission in ^{235}U, but do not cause fission in, and are not captured by, ^{238}U. A chain reaction is thus possible in an array of fuel elements embedded in the moderator, as shown schematically in figure 8.1. The sequence of events is shown in figure 8.2.

A reactor which operates on the principles described above is called a *thermal reactor.* Nearly all the reactors constructed to date have been of this type, using as fuel either natural uranium or uranium slightly enriched (to a few per cent) in ^{235}U. The power reactors producing electricity for the national grid in the U.K. have graphite moderators and the heat is removed by flowing CO_2 gas. Various other types of reactor have also been built.

A different type of reactor being considered for power generation is the *fast reactor,* which makes use of fission by neutrons which have undergone relatively little slowing down by collisions. Such a reactor uses no moderator, but in order to sustain a chain reaction a high

150

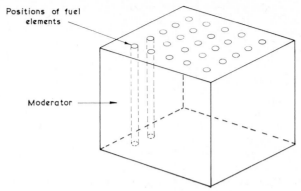

Fig. 8.1. Schematic diagram of the arrangement of fuel elements in the moderator of a thermal fission reactor.

proportion of the fuel must be in a form which is fissionable by neutrons in a wide energy range. If uranium is used as a fuel, this means that it must be highly enriched (>20 per cent) in ^{235}U. An alternative fuel is ^{239}Pu which, as noted before, is produced as a by-product of thermal reactors by neutron capture in ^{238}U. The fast reactor is thus only feasible if such fuels are available, but it has an advantage over thermal reactors because enough fast neutrons are available from each fission event not only to sustain the chain reaction but also to convert ^{238}U

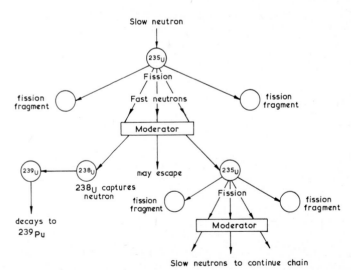

Fig. 8.2. A representation of the sequence of events in a fission nuclear reactor (after S. Glasstone, *Sourcebook on Atomic Energy,* Van Nostrand Inc., Princeton, New Jersey, 1967).

151

into ^{239}Pu as quickly as, or more quickly than, the fissile fuel is being consumed. This is known as breeding, and one of the main reasons for developing the fast reactor is that such a system could make much more efficient use of natural resources of uranium than can a thermal reactor, which consumes the ^{235}U and only converts a small proportion of the highly abundant ^{238}U into ^{239}Pu.

From the discussion above it can be seen that the very nature of the fission process means that a reactor is subject to radiation damage problems. The types of radiation present in the reactor may be summarized as follows:

(i) Fission products. These are heavy ions with relative atomic mass between ~80 and ~160, and having initial energies up to ~100 MeV. They account for nearly all the energy released in the fission process. However, the range of such a projectile is only a few tens of micrometres (see Chapter 6) and as a result only the fuel itself suffers serious damage from fission fragments.

(ii) Neutrons. These are emitted during the fission process (a few neutrons per fission event) with energies from ~0·1 MeV up to ~10 MeV, most having energies in the 1–2 MeV region. These are then slowed down by collisions with nuclei in the components of the reactor, particularly the moderator in a thermal reactor. The spectrum of neutron energies thus depends on the type of reactor and the position in the reactor. In a thermal reactor the neutron energies range from ~0·05 eV up to ~10 MeV, nearly nine decades in energy. Some typical neutron spectra in reactors are shown in figure 8.3 and fluxes are given in table 8.1. Generally speaking, only neutrons with energies $\gtrsim 0·1$ MeV are significant for producing radiation damage by direct momentum

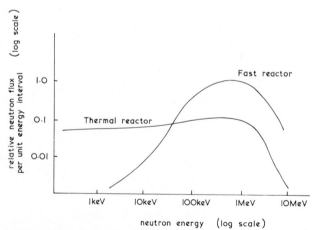

Fig. 8.3. Typical energy distributions of the neutrons in thermal and fast reactors.

152

transfer processes (Chapter 7), and it is customary to refer to them as 'fast neutrons'. (Very often, however, a quoted fast neutron dose refers to energies >1 MeV). Low energy neutrons may produce some radiation damage through nuclear relations (see Section 7.4). Since they have long ranges, fast neutrons are responsible for nearly all the radiation damage to components of the reactor outside the actual fuel itself.

Table 8.1. Typical neutron fluxes and displacement rates near the fuel elements in nuclear reactors.

	Fast neutron flux neutrons $m^{-2} s^{-1}$	Displacement rate atoms $m^{-3} s^{-1}$	Displacements per atom in one year
Thermal reactor	10^{16}–10^{17}	10^{19}–10^{20}	10^{-2}–10^{-1}
Fast reactor	10^{19}–10^{20}	10^{22}–10^{23}	10–100

(iii) γ-rays. The total energy released in γ-rays during the fission process itself and subsequent radioactive decays of fission products is comparable with that of the fission neutrons (about 10 MeV per fission). Since γ-rays are highly penetrating, all components of the reactor are subject to their irradiation. In most solids γ-rays are in-effective in producing radiation damage, and the main effect of the γ-rays in the reactor is heating. However, if a material which shows photochemical damage (Section 7.3) is placed in a reactor, the damage it sustains through the γ-rays will be at least as important as the damage from fast neutrons.

(iv) Electrons and other particles. Some electrons and positrons with initial energies of about 1 MeV are present in the reactor as a result of decay of fission products. These will be accompanied by neutrinos. The electrons, positrons and neutrinos account for about 15 MeV per fission. Since electrons are much less effective at producing atomic dis-placement than neutrons and have short ranges (few millimetres), their contribution to reactor radiation damage can be ignored.

A final point of considerable importance to radiation damage events in reactors is that the temperature is high. The temperature in the core of a gas-cooled thermal reactor as used in the first generation of nuclear power stations is about 400°C, whilst future power reactors will have even higher operating temperatures (to increase the temperature of the steam generated and hence the efficiency of the turbines). Secondary radiation-damage effects, as discussed in Section 7.5, thus take on great importance at these temperatures where primary point defects are mobile.

In addition to the fast neutron fluxes, table 8.1 also gives some typical displacement rates for the atoms in a solid subject to the neutron flux, and the number of displacements per atom in a period of one year. It can be seen that a very high level of radiation damage is likely

153

to be sustained by any reactor component, especially in a fast reactor with its high fast neutron flux.

Fusion reactors

The sun and stars generate energy by the process of nuclear *fusion*. This is the fusing together of the nuclei of light elements to form a stable nucleus, energy being released in the process. A fusion reaction is also used in the hydrogen bomb. For many years scientists have been trying to find conditions under which such a reaction could proceed in a controllable way on earth, and thus provide an alternative source of nuclear power to the fission reactor. The great advantage of such a *fusion reactor* would be the abundance of its fuel. The most favoured fusion reaction is the so-called D–T reaction:

$$^2H + ^3H \rightarrow ^4He + ^1n + 17 \cdot 6 \text{ MeV}$$

which provides a 14·1 MeV neutron and a 3·5 MeV α-particle. Unlike the fission process, most of the energy released in fusion resides in the energy of the neutrons. The tritium for the reaction would be obtained from the following reactions involving lithium and a neutron:

$$^1n + ^6Li \rightarrow ^4He + ^3H + 4 \cdot 79 \text{ MeV}$$

$$^1n + ^7Li + 2 \cdot 5 \text{ MeV} \rightarrow ^4He + ^3H + ^1n$$

Both deuterium and lithium are very abundant, so the prospect of power from fusion is very attractive. The problem is to achieve the conditions under which fusion may occur in a stable and controlled way (a high density of atoms at very high temperatures, often called a plasma, is necessary). However, a great deal of progress has been made towards this goal, and it now seems possible that fusion reactors will be with us in the next century.

Fusion reactors, however, will not be without their problems. With the designs envisaged, the flux of 14 MeV neutrons will be $>10^{19} \text{ m}^{-2} \text{ s}^{-1}$, so the radiation doses will be comparable with those in a fast reactor. Particular new problems are presented, however, which will be discussed in Section 8.2.

8.2. Radiation damage in reactors

Stored energy

E. P. Wigner was amongst the first to recognize that nuclear reactor components would suffer degradation in their physical properties as a result of radiation damage. One of the earliest examples of this was the problem of stored energy in the graphite used as a moderator. When defects are created in a solid by irradiation, the material is placed in a metastable state. The defect-containing solid has more energy than a perfect solid, and this energy will be released if the solid is heated to a temperature at which defect annealing occurs. At high doses the stored energy can be appreciable. Consider a solid containing 10^{26} m^{-3} vacancy–interstitial pairs, which is one defect every $\sim 10^3$ lattice sites. Suppose the energy per pair is 10 eV, a reasonable figure (see Chapter 3). The total stored energy is thus

$$U = 10 \times 10^{26} \text{ eV m}^{-3} = 1 \cdot 6 \times 10^8 \text{ J m}^{-3}$$

Since the heat capacity of most solids is $\sim 10^6$ J m^{-3} K^{-1}, if all this energy were released it could raise the temperature of the solid by about a hundred degrees. In graphite stored energy of up to $\sim 5 \times 10^9$ J m^{-3} has been observed after fast neutron irradiation to doses of a few times 10^{24} neutrons m^{-2}, which could obviously be serious if the temperature of operation of the reactor covered the range over which annealing processes take place and stored energy is released. In graphite there is a stored energy release peak at about 200°C known as the 'Wigner release peak', and a further peak near 1300°C. Early graphite-moderated reactors operated with the moderator at a mean temperature of ~ 100°C, at which temperature stable radiation damage accumulates, causing dimensional changes (see Section 4.1 and later in this section). Since these cause distortion of the structure, it was desirable to anneal them out periodically by heating the moderator in excess of 200°C. Great care was required for this operation because the release of sizeable amounts of stored energy could cause the system to go out of control. Indeed, an accident of this type occurred at Windscale in the 1950's, during which overheating occurred and radio-active fission products were accidentally released into the atmosphere. Later generations of graphite-moderated reactors have operating temperatures above 200°C, and the problem of accumulated damage and stored energy release is reduced.

Swelling

In Chapter 4 we have already mentioned dimensional changes caused by defects. The examples we considered there, however, were fairly fundamental ones of point defects causing changes in macroscopic volume and lattice parameter. In reactor environments the radiation damage occurs at elevated temperatures, and the stable radiation-damage products produced are generally not point defects, but larger aggregates which form as a result of point defect diffusion. Also, as we have seen in table 8.1, the fast neutron doses suffered by a reactor component can be very high, so that large amounts of damage are sustained.

Graphite is one of the most widely studied reactor materials, as a result of its use as a moderator. The crystal structure of graphite consists of layers of carbon atoms, the arrangement of the atoms in each layer being hexagonal. This layer structure gives a single crystal of graphite highly anisotropic properties, and indeed, irradiated graphite expands along the directions perpendicular to the layers and contracts in directions in the plane of the layers. The expansion can be very high at large doses ($\sim 10^{24}$ neutrons m^{-2}) and fractional length changes larger than 10 per cent have been observed. Both the lattice parameter and macroscopic length increase, but the fractional change in the latter is larger than in the former. The explanation for this is indicated in figure 8.4. The swelling is thought to be due to interstitial clusters, some

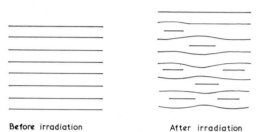

Before irradiation   After irradiation

Fig. 8.4. Illustration of swelling in irradiated graphite caused by interstitial dislocation loops lying parallel to the layers of the crystallographic planes. The vacancies are distributed as small clusters, which are not shown.

of which are in the form of large dislocation loops. Vacancies are much less mobile than interstitials and remain isolated or in small clusters at moderate temperatures ($\lesssim 500°C$). The interstitial loops form new layers in the graphite crystal structure. Such large loops cause greater length change than lattice parameter change, since they represent new crystallographic planes in the lattice. The lattice parameter changes are small because lattice strains are only present near the boundary of a loop, which represents a small proportion of its area. The loops may also be observed directly in the electron microscope (Chapter 5). The graphite in a reactor moderator is polycrystalline, so the changes in dimension caused by irradiation are a superposition of the effects in the different crystal grains, and can be quite complex.

Some remarkable dimensional changes occur in uranium metal, which if used as the fuel in a nuclear reactor is exposed to very high doses of radiation from both the fission neutrons and the fission fragments. The stable form of uranium at moderate temperatures is α-uranium, which has an orthorhombic crystal structure. (This is the same symmetry as a box with rectangular sides; a single crystal will have three unequal lattice spacings along three orthogonal directions). When heavily irradiated, α-uranium expands along one direction, shrinks along one orthogonal direction and does not change in dimensions along the third direction. The result is shown in figure 8.5. These highly directional dimensional changes are thought to be due, again, to the formation of dislocation loops preferentially on certain crystallographic planes. These

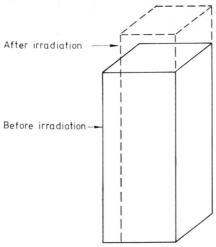

After irradiation

Before irradiation

Fig. 8.5. Schematic illustration of swelling in an irradiated single crystal of α-uranium.

length changes are accompanied by very little *volume* change, but operation in a reactor at high temperatures ($\sim 400°C$) can cause very large volume changes of uranium metal and its compounds because of the evolution of gaseous fission products. About 10 per cent of the fission fragments produced are the rare gases Xe and Kr, which precipitate into bubbles if the temperature is high enough for them to diffuse in the lattice, thus causing swelling. In equilibrium the pressure excess in the bubble balances the surface energy, that is $\Delta p = 2\gamma/r$, where r is the bubble radius (a simple surface tension formula). Some of the gas also escapes from the fuel. Since these fission products may be radioactive, they must not be allowed to leak out of the reactor. The fuel elements used in a reactor are therefore not simply composed of the fuel itself, but consist of the fuel contained in a can (cladding), which is designed to prevent escape of fission products, as well as to contain the fuel in as rigid an environment as possible. The first generation of British power reactors (Calder Hall, Wylfa, and others) use uranium metal fuel with magnesium alloy (magnox) cladding, whereas the advanced gas-cooled reactors presently being constructed use UO_2 fuel clad in stainless steel. Both of these features allow higher operating temperatures for the reactor than is possible in the first generation systems. The cladding itself, of course, as well as all other components in a reactor, undergoes radiation damage from the fast neutrons.

A problem of considerable technical importance which has come to light in the past few years is that of voids in fast reactor components, particularly the cladding of fuel elements. The components in a fast reactor are subject to very high doses of fast neutrons at operating temperatures of 400°–600°C (that is, up to about half the melting temperature of stainless steel). It had been thought that this would cause no extra substantial radiation-damage problems, because the expected stable defects at these temperatures were vacancy and interstitial dislocation loops (see Chapter 3). Since these cause opposite changes in macroscopic dimensions, little swelling was expected. However, in 1966 it was found that samples taken from the cladding of fuel elements in the Dounreay fast reactor irradiated to neutron doses greater than 10^{26} neutrons m^{-2} showed nearly spherical voids up to ~ 100 nm in diameter, and swelling of several per cent. The reason for the swelling is that a given number of vacancies in a spherical void causes a very much smaller contraction in the solid than when dispersed in vacancy dislocation loops. In view of its importance this problem has been studied widely since its discovery, using samples irradiated for long periods in fast reactors and also by simulating reactor damage using heavy ion and electron irradiation for shorter periods (see Chapter 7). Figure 8.6 shows electron micrographs of voids in nickel, designed to show the equivalence of fast reactor and heavy ion irradiations. Figure 8.7 shows the temperature and dose dependence of void-induced swelling in steel. It is worth noting (and thinking about!) that the swelling is caused not by the voids themselves, but by their complementary interstitials. (Making holes in a sample does not change its macroscopic volume, but finding room for the removed material does). Void formation was unexpected, but the reasons for its occurrence are now thought to be understood. The explanation lies in the complexity of stable defect formation under high temperature conditions in which both vacancies and interstitials are highly mobile. Two questions are raised: why do voids form in the first place instead of vacancy dislocation loops, and how do they grow? The answer to the first question seems to be that the void configuration is stabilized by the presence of gas atoms which are relatively insoluble in the material being irradiated. In particular, helium is produced during neutron irradiation by (n, α) reactions, and there is evidence from heavy ion and electron irradiation experiments that some gas atoms must be present in order for voids to form. These gas atoms collect in small clusters of vacancies produced during the early stages of irradiation. Once a vacancy cluster contains some gas it is not energetically favourable for it to collapse into a dislocation loop (normally a lower energy configuration because defective lattice only exists at the core of the dislocation), and instead a void is nucleated. The void, however, is not a gas bubble in the sense previously described, in that the pressure of gas is not itself sufficient to balance the surface energy. The *growth* of the void requires that vacancies may be added without this being balanced by an equal flow of interstitials (which would, of course, annihilate an equal number of

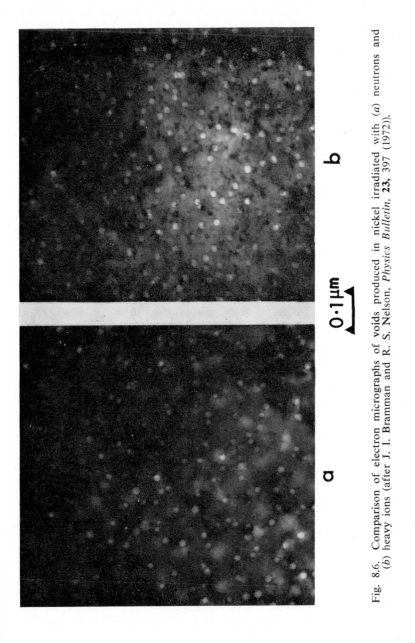

0·1 μm

Fig. 8.6. Comparison of electron micrographs of voids produced in nickel irradiated with (*a*) neutrons and (*b*) heavy ions (after J. I. Bramman and R. S. Nelson, *Physics Bulletin*, **23**, 397 (1972)).

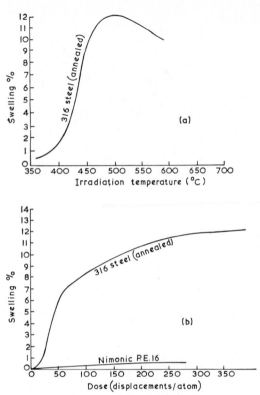

Fig. 8.7. (a) Void swelling as a function of temperature in stainless steel. (b) Swelling versus irradiation dose for stainless steel, and a Nimonic alloy (after J. I. Bramman and R. S. Nelson, *Physics Bulletin*, **23,** 397 (1972)).

vacancies). It is now generally accepted that void growth is possible because dislocations are slightly more effective as sinks for interstitials than for vacancies. As a result, some vacancies are available to feed the voids, without annihilation occurring. Void growth only occurs at temperatures high enough for vacancies to have high mobility, as may be appreciated from figure 8.7. At very high temperatures thermally generated defects (see Section 3.2) become more important than radiation-induced defects and the voids become unstable. Some of the basic knowledge accumulated about voids has enabled materials to be chosen which show reduced void formation. Figure 8.7 shows one example, a nickel-based precipitation-hardened alloy ('Nimonic').

Embrittlement and creep

The mechanical integrity of reactor components is not limited to resistance to swelling, but extends to their behaviour under conditions of stress at high temperatures. The stress, of course, may arise because of dimensional changes elsewhere! It was pointed out in Chapter 4 that plastic flow in ductile solids could be reduced by the introduction of defects which inhibit dislocation motion, but that if this process is pressed too far, then the solid will become brittle. The radiation-induced defects in reactor components can have this undesirable effect, and care must be taken in reactor design to ensure that vital components will not undergo embrittlement and fracture during operation.

Creep of components under stress at high temperatures is caused by defect motion (such as dislocation climb), and can obviously also be affected by irradiation. For example, the trapping of interstitials at dislocations mentioned in connection with void growth will cause dislocation climb, and therefore creep of a component under stress. Figure 8.8 shows a radiograph of a loaded beam of material before and after it has undergone test irradiation in a reactor. It can be seen in the right-hand photograph of figure 8.8 (b) that significant movement has occurred. The use of high energy ions to simulate reactor irradiation with neutrons is again useful in creep measurements, because in-reactor tests are difficult to make.

Damage in fusion reactors

The neutron fluxes and operating temperatures of components envisaged for future fusion reactors are similar to those in the fast fission reactor, so that similar problems apply. The higher initial energy of the neutrons (14 MeV) means, however, that damage rates and production rates of some gases by nuclear reactions are increased for the same neutron flux. One problem which does not occur in a fission reactor arises from the fact that a magnetic field is required to hold the very hot reacting fuel (the plasma) in a stable condition away from the walls of the containment vessel (otherwise the necessary high temperature could not be achieved). It is very likely that this magnetic field will be generated by a superconducting magnet, which will surround the containment vessel, and, incidentally, will form the largest single cost item in the whole reactor. A proposed structure is indicated in figure 8.9, which shows the cross section of the toroid which forms the fusion reactor configuration. Although some shielding of the magnet will be possible, if this becomes too large then the size and cost of the magnet (which requires cooling with expensive liquid helium) will rise. There will thus inevitably be some neutron irradiation of the magnet material which could lead to a degradation in its performance. Experiments are presently under way in many parts of the world to examine the effects of irradiation on superconductors irradiated at low temperatures (<20 K). Some effects which may occur are:

161

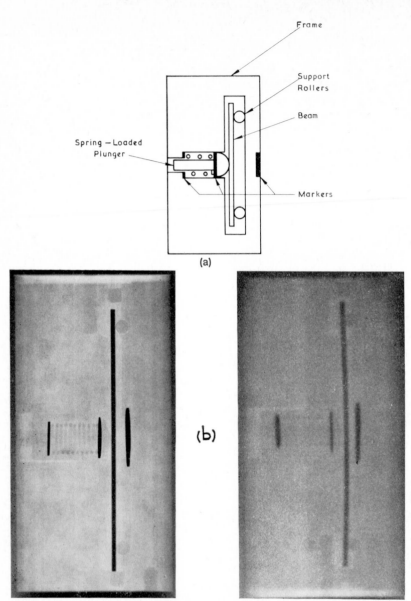

Fig. 8.8. Creep in an irradiated material. (a) shows the experimental arrangement used to apply a bending stress to a beam of material. (b) shows a neutron radiograph of the arrangement before irradiation (left-hand photograph) and after irradiation (right hand photograph) in a nuclear reactor. The poorer quality of the right hand photograph is a result of radioactivity induced by irradiation (courtesy of the Harwell Neutron Radiography Service).

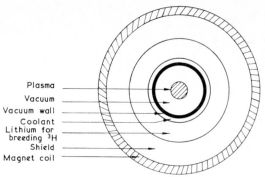

Plasma
Vacuum
Vacuum wall
Coolant
Lithium for breeding ^{3}H
Shield
Magnet coil

Fig. 8.9. A section through the toroidal containment ring of an envisaged fusion reactor.

(i) Defects may reduce the 'hardness' of the superconductor, which is its ability to remain superconducting in the presence of a magnetic field.

(ii) Compound superconductors such as niobium tin (Nb_3Sn), which are capable of providing the highest fields, may become disordered alloys rather than ordered ones as a result of bombardment. (An ordered alloy is one in which the two components occur in definite order on the crystal lattice rather than randomly.)

These studies and assessment of other radiation-damage problems form an important part of planning for the fusion reactor of the future.

8.3. *Radiation effects in space*

The past ten years or so have seen an enormous increase in the number of satellites orbiting the earth, for purposes ranging from communications to space exploration. All these satellites contain a wide variety of electronic components, mostly solid-state devices such as transistors, all of which are exposed to a greater or lesser degree to the radiation belts around the earth. Typical fluxes encountered are $\sim 10^{10}$ m^{-2} s^{-1} of 1–2 MeV electrons and $\sim 10^{11}$ m^{-2} s^{-1} of 200–500 keV protons. Of course, studies of radiation effects on electronic devices started before satellites went into orbit, since instrumentation is required in reactors and in other potentially hazardous radiation environments (for example, military and civilian equipment may be exposed to the aftermath of a nuclear explosion). However, the increasing number and sophistication of solid-state devices and the special needs of space applications have ensured that this remains a large field of research activity. We shall mention here two examples of the importance of radiation damage in electronic components. Radiation also has import-

163

ant effects on many of the other materials used in space, for example, coatings for controlling the temperature of surfaces exposed to the sun, and the structural members of satellites.

Solar cells

Solar cells are crucial in providing electrical power for satellites and space vehicles. A typical solar cell structure is shown in figure 8.10. It

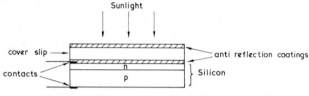

Fig. 8.10.   The structure of a silicon solar cell.

consists of a p–n junction in silicon about $0.25\,\mu$m below the front surface of the cell. Light falling on the cell generates electrons and holes, mainly in the p-type region behind the junction. Some of the 'minority' carriers (electrons in the p-type region) diffuse towards the p–n junction and are swept into the n-type region by the internal electric field at the junction (see Chapter 4, figure 4.5). As a result, a voltage is produced across the device (a photovoltaic effect). The energy conversion efficiency is 10–15 per cent. The power produced by the cell depends on the efficiency with which the electrons can reach the junction region after being generated in the bulk of the p-type silicon. Radiation-induced defects reduce the efficiency by acting as traps and recombination centres for the carriers. As a result the output of the cell decreases with increasing dose as shown in figure 8.11. Since a

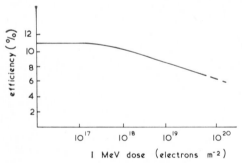

Fig. 8.11.   The degradation in the efficiency of a solar cell during irradiation with 1 MeV electrons.

satellite may be required to spend many years in orbit, a rapid degradation is intolerable, and considerable efforts have been put into examining and improving the radiation resistance of solar cells. Any improvement in degradation characteristics essentially means a weight saving on the satellite, since then spare capacity requirements are reduced.

Low energy protons (a few hundred keV) are particularly harmful to the cell because their range is comparable with the depth of the p–n junction. Defects produced in the junction region have a short-circuiting effect on the cell. For this reason (and to improve the emissivity of the cell for cooling reasons) all solar cells are fitted with a 'cover-slip' which usually consists of a glass slide about 100 μm thick. Various anti-reflection coatings are also employed to maximize the light intensity at the cell itself. The cover-slip protects the cell from low energy protons (and partially from other radiation), but may itself suffer damage. This is important if colour centres are produced which absorb some of the light otherwise destined for the cell. In practice various cerium-loaded glasses or silica are found to be adequate in this respect. Nevertheless, it can be seen that the protective devices increase the weight of the cell, and the presence of radiation in space adds to the design problems of satellites.

Charge-trapping effects in devices

Devices called metal-oxide-silicon field effect transistors (MOSFETs) are widely used in modern electronic circuits as 'gates', that is, they conduct if a certain voltage threshold is exceeded. The construction of one type of MOSFET and its current–voltage characteristic are shown in figure 8.12. Irradiation is found to change the characteristics of the device, in particular the threshold voltage is changed as indicated in figure 8.12. This would obviously upset its use in electronic circuitry.

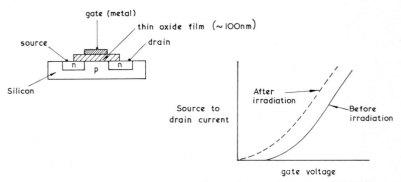

Fig. 8.12. Structure and current–voltage characteristic of a MOSFET.

The effect is thought to be due to the generation of electrons and holes in the insulating oxide of the gate. The electrons are mobile and can be swept out of the oxide when a voltage is applied, but the holes become trapped and consequently give rise to a positive charge in the

165

oxide. This trapped positive charge then changes the characteristic of the MOSFET. For example, in the case shown in figure 8.12 the positive charge simulates an extra positive voltage on the gate. The magnitude of the effect is found to depend on the voltage applied to the gate during irradiation, as might be expected from the above explanation. Some successes have been achieved over the past few years in minimizing the radiation effects and producing more radiation-resistant ('hard') devices, but the basic mechanisms are still not very well understood.

8.4. *Miscellaneous problems*

Phosphor degradation

Finally, we make brief mention of some miscellaneous effects which may be classed as radiation-damage problems of technical importance. One of the most widely used electronic devices is the cathode-ray tube; it appears in nearly every home as part of a television set, and in every laboratory as part of an oscilloscope. A crucial component of the tube is the phosphor, which emits light when struck by electrons, usually having energies of 10–20 keV. The main function of the electron beam is to generate electron–hole pairs in the phosphor material, which then recombine (usually at impurity sites) and emit light (see Chapters 5 and 7). The radiation dose received by the phosphor can, however, be very large, as may be seen by the following crude calculation for a stationary spot on, say, a cathode-ray oscilloscope.

A typical electron beam current is $\sim 1\ \mu A$, which falls in an area of $\sim 1\ mm^2$ i.e. $10^{-6}\ m^2$. The incident power on the phosphor is therefore, for a voltage of 10 kV,

$$P = \frac{10^{-6} \times 10^4}{10^{-6}} = 10^4\ W\ m^{-2}$$

Since the range of a 10 keV electron in most solids is about 1 μm, the volume power density P_v is

$$P_v = \frac{P}{10^{-6}} = 10^{10}\ W\ m^{-3}$$

If the tube is operated on one spot for 10^3 s the total dose delivered to the phosphor is

$$D = P_v \times 10^3 = 10^{13}\ J\ m^{-3}$$

or about 10^3 eV for each atom in the spot. Thus even if there is only a relatively inefficient photochemical radiation-damage process in the phosphor, some defects will be produced. For example, if 10^6 eV is required to make one defect, about 10^{26} defects m^{-3} will be produced, which is comparable with the concentration of impurities responsible for emitting the light. In such a case the defects may well upset the luminescence process and a degradation of the phosphor efficiency will result.

Fortunately the phosphors used in television screens (usually mostly ZnS, with Y_2O_3 in colour tubes) do not suffer from this effect, but specialized phosphors used in radar screens (where long persistence of the image is required) often do. It is very difficult to get round this problem, except by choosing a different phosphor which does not degrade. Sometimes, however, the choice is limited by other factors such as brightness and some degradation must be tolerated.

Lasers

A completely different radiation-damage problem from those we have discussed so far is the ability of lasers to damage materials and themselves. Lasers produce very intense beams of coherent radiation in the visible and near-visible regions of the spectrum, many lasers using solid-state materials (for example, ruby lasers and gallium arsenide semiconductor lasers). These intense beams of radiation may cause physical damage to solids, and indeed to the laser material itself. The mechanisms of damage are generally not well understood, except where they involve direct heating of material by the laser beam with consequent thermal shock, melting or evaporation. Since some types of laser may find widespread application in future communications systems, some of these damage problems, especially where self-damage is involved, are becoming of increasing importance.

In this chapter we have seen how radiation damage to structural materials and electronic components is a widespread effect which must be taken into account by design engineers in a wide variety of situations. Although the basic processes of primary defect production, especially by momentum transfer processes, are reasonably well understood, the conditions demanded in practice are often sufficiently complex that detailed predictions about material performance cannot be made adequately. There thus remains much scope for radiation-damage studies, especially as new technologies (such as the fusion reactor) emerge with their own special requirements. All the examples we have discussed here have been problems: in other words, life would be much easier if they did not exist. In the next and final chapter of the book we take a look at some more positive aspects of radiation effects, in which the properties of radiation-induced defects are used to produce useful and desirable changes in solids.

CHAPTER 9

applications of radiation effects

Having looked at the many unfortunate consequences of radiation damage, in fairness we must also consider what uses radiation-induced defects find. In fact we will see that these are also many in number, and growing in importance. They can conveniently be classified into three groups. In the first group, radiation dosimetry, the creation of defects is used as a convenient way of measuring radiation dose integrated over long periods of time. In the second group, information storage devices, the defects themselves are made the analogue of stored information, as in photographic film, and in most cases the host material properties are of minor importance. In the final group we consider applications where the radiation is used to change the bulk properties of a material in some desirable way.

9.1. *Solid state radiation dosimetry*

One characteristic of our age is its rapidly increasing use of high energy radiation. X-rays are already widely used for radiography in medicine and industry, for diagnosing tumours, bone fractures, cracks in steel structures, etc. They are also used for sterilization of medical supplies and for treatment of certain diseases. High voltage electron beams are beginning to be used for welding and paint curing. Perhaps the most obvious sources are the nuclear reactors now widely used for electrical power generation, although in practice they are so carefully controlled that there is almost no radiation outside the reactor structure. Nevertheless a large number of people are in potential contact with high energy radiation, and it is essential that their contact with it should not be excessive.

The exposure of living things to radiation is usually measured in terms of the received *dose* by a radiation *dosimeter*. Roughly speaking the dose is proportional to the energy deposited per gram of tissue, and a convenient unit is the rad, which corresponds to a deposited energy of 10 mJ kg^{-1}. The law requires that the radiation dose received by everyone likely to be in contact with radiation be regularly and continuously monitored. How can it be done cheaply and conveniently?

In Chapter 6 we saw how photons and charged particles lose most of their energy to the electrons of the atoms with which they interact. It is therefore not surprising that all sensitive radiation detection processes depend on the excitation of electrons. For example, intrinsic semiconductors may be used; these are semiconductors free from electrically active impurities and which are therefore good insulators at room temperature. When a high energy photon is absorbed by such a crystal a large number of electron–hole pairs is created, and the crystal becomes a conductor for a very short time (about $1\,\mu s$). If a voltage is applied at this time a current pulse flows and the charge flow is proportional to the energy of the photon absorbed. If the pulses of charge caused by a large number of photons are added together with a suitable electronic circuit, then the total charge which flows in a given time is proportional to the radiation dose received by the crystal. A detector of this kind can be carried by a radiation worker to monitor his received dose, but it has a number of serious disadvantages. The most obvious is that it is fairly bulky, expensive and delicate, since it contains an expensive single crystal, a battery and some electronic circuits. It also has to include a memory of some kind, usually a mechanical counter, in which the record of the total charge flow is kept.

Less obvious, but a serious disadvantage all the same, is the fact that even silicon has an atomic number significantly different from the mean atomic number of human tissue. In Chapter 6 we saw that the cross-sections for photo-electric absorption, Compton scattering and pair-production varied with Z^4, Z and Z^2 respectively. It is therefore impossible to ensure that the dose received in human tissue from an arbitrary photon radiation field is proportional to that received in a dosimeter unless the Z of the dosimeter material is very nearly the same as that of tissue. Solid-state integrating dosimeters aim to solve all the problems we have listed, and succeed to some extent. Thus materials are used where electronic excitation events (hence high sensitivity) lead to permanent changes (so that external memory is not necessary). These changes can be detected readily at the end of the monitoring period, and always materials are chosen with Z near 8, since the radiation sensitivity of human tissue is usually dominated by the oxygen content of water.

The most widely used personnel dosimetry material at this time is still photographic film. The silver-halide film is wrapped in a light-proof paper and is therefore sensitive only to radiation which can penetrate this cover. The high energy radiation which does penetrate the wrapping exposes the film in just the same way as light would. Electron–hole pairs are created in the silver-halide crystals, which are dispersed in a gelatine matrix. The electrons and holes are trapped to form the sensitive points (the latent image) where metallic silver grain growth can begin during chemical development. At the end of the radiation monitoring period the film is developed and the blackening is proportional to the received dose. Photographic film dosimeters are certainly cheap and sensitive, but they do have the disadvantage of requiring complex processing, which is difficult to control precisely, and

169

they do respond to radiation of different types very differently from tissue, because the active material is a silver-halide crystal with average Z about 40. This latter problem is usually dealt with by mounting the film in a holder between a complex set of radiation filters, which allow an estimate to be made of how tissue would respond to a particular kind of radiation from measurements of how each film filter segment responds, but this technique is complex and not very satisfactory.

Two different types of integrating personnel dosimeter have therefore been developed, with the objective of eliminating complex processing and the need for extensive Z correction. These use photoluminescence and thermoluminescence processes respectively. In both cases luminescent centres are created by radiation and the two techniques have a rapidly growing importance.

Radiophotoluminescence

In this kind of dosimetry, materials are chosen in which defects are created by radiation (hence 'radio') suitable for luminescence excited by absorbed photons (hence photoluminescence).

The materials most widely used are phosphate glasses containing several per cent of silver ions. A typical recipe uses a glass base made from 47 per cent $LiPO_3$ and 53 per cent $Al(PO_3)_3$ to which about 12 per cent of $AgPO_3$ is added. These glasses are normally transparent to light of wavelength longer than about 300 nm, but on irradiation defects are created which have a pronounced absorption band at around 320 nm (figure 9.1). If the irradiated glasses are excited with light of 320 nm wavelength a strong orange luminescence results, centred on 640 nm, whereas no luminescence in this region is produced by unirradiated glasses.

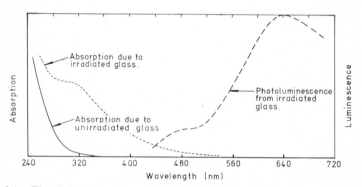

Fig. 9.1. The behaviour of radiophotoluminescent silver-activated phosphate glasses (after J. H. Schulman, *Proceedings of the International Conference on Luminescence Dosimetry, U.S.A.E.C.* C-650637 (1967)).

Both the absorption band at 320 nm and the emission band at 640 nm can be, and are, used as a measure of the radiation dose received by the sample, but for a number of reasons luminescence is the more attractive in most applications. These advantages of luminescence are not, of course, peculiar to phosphate glasses but apply to any similar dosimeter material. The primary advantage of luminescence is its sensitivity. Personnel dosimeters are required to detect doses down to about 10 milli rad (6×10^{14} eV kg^{-1}), and in a sample about 10 mm $\times$ 10 mm $\times$ 1 mm such a dose would create at most $\sim 10^{10}$ defects, even by an efficient electronic or photochemical process. If these defects were strong absorbers like F centres in alkali halides the sample would still absorb only about 10^{-5} of light incident at the absorption band wavelength, and in order to detect such absorption the transmitted light would have to be measured with an accuracy of five significant figures, which is exceedingly difficult. If, on the other hand, the defects were to luminesce efficiently, then an exciting beam of 10^{14} photons s^{-1} (about 100 μW) would cause 10^9 ($10^{-5} \times 10^{14}$) defect excitations s^{-1} and up to 10^9 photons s^{-1} would be emitted, depending on the efficiency of the luminescence process. These luminescence photons would have a longer wavelength than those used for excitation and the two could therefore be separated very easily. The problem of distinguishing between zero dose and a dose of 10 milli rad by photoluminescence therefore reduces to distinguishing between zero luminescence and an output photon flux of 10^9 s^{-1}, and this is fairly easy if the exciting flux of 10^{14} s^{-1} can be separated by a monochromator. Using absorption, remember, we had to distinguish between $1 \cdot 00000 \times 10^{14}$ photons s^{-1} and $0 \cdot 99999 \times 10^{14}$ photons s^{-1}.

A second advantage of luminescence is its relative insensitivity to the physical state of the sample. For example, powdered, and therefore cheap, material can be used, whereas high optical quality materials are necessary for absorption. It is not uncommon for the fractional optical transmission of a sample to vary by a few per cent with age, primarily caused by surface degradation. In our hypothetical dosimeter a dose of 1 rad caused only $0 \cdot 1$ per cent change in transmission coefficient so that an accidental change of say 3 per cent would be interpreted as a dose of 30 rad by an absorption dosimeter. On the other hand, the luminescence dosimeter suffering the same changes would be only 3 per cent in error, that is, it would read not 10 milli rad but $9 \cdot 7$ milli rad, and this difference is very rarely important.

Radiophotoluminescent dosimeters are therefore quite sensitive; they can be made cheaply and they avoid the need for the complex and unreliable processing necessary with photographic film. Since the active material is Li, Al, P and O with a little Ag the mean atomic number and therefore the sensitivity to different kinds of irradiation is much more like human tissue than is film, where the active materials are AgCl and AgBr. They can detect doses down to about 10 milli rad and have a response linear with dose up to about 1 kilo rad. The absorption band at 320 nm can be used to measure even higher doses. These dosimeters were developed originally for the U.S. Navy and have been widely used in the U.S., but they are less popular generally than thermoluminescence dosimeters, for several reasons which we will outline later.

The way in which radiophotoluminescence dosimeters function has been deduced from work on silver activated alkali halides, which often behave in a similar way but are much easier to study than the amorphous glasses. Using KCl doped with Ag^+ ions it has been shown that Ag^+ ions act as traps for holes, forming Ag^{2+} ions during irradiation. Pairs of Ag^+ ions, to which chlorine ion vacancies have probably been attached, act as electron traps and during irradiation form complex defects of the type shown in figure 9.2. These latter defects, the ones with trapped electrons, have an absorption band at 310 nm and an emission band at 556 nm in KCl. They therefore behave like the photoluminescence centres in the phosphate glasses and it has been assumed that the centres in the glasses are like those in figure 9.2.

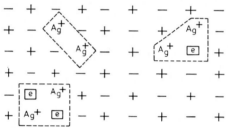

Fig. 9.2. The atomic structure of photoluminescent centres created by irradiation in KCl : Ag.

Radiothermoluminescence

Whenever radiation creates defects in a solid, energy is stored. Very often the defects are stable for long periods of time at room temperature, but if the crystal is heated electrons or holes begin to be released or the defects themselves begin to diffuse or both. When this happens the energy stored is released, usually as heat, as we discussed in Chapter 8, but sometimes partly as light. The latter phenomenon is called thermoluminscence, because the material luminesces when heated, and is the basis for the most promising form of dosimetry known at present.

The general advantages of thermoluminescence are many. As we saw before, we expect a dose of 10 milli rad to create about 10^{17} defects m^{-3}, so that a convenient sample of material about $2 \cdot 5 \, mm \times 2 \, mm \times 2 \, mm$ would contain 10^9 defects. If as few as 1 per cent of these defects caused luminescence events on subsequent heating there would be an output of 10^7 photons per 10 milli rad dose, which can be detected quite easily. Thermoluminescence is therefore very sensitive. It scores over photoluminescence in that no light source and monochromator system is necessary for 'reading' the dosimeter, but instead a very simple heating system with a photomultiplier detector can be used. To a first approximation the number of photons emitted by a thermoluminescence dosimeter depends only on the dose it has received and not, for example, on the heating rate used during reading, which makes reader-calibration fairly easy. Yet another advantage of thermoluminescence is that reading the dosimeter automatically makes it ready for re-use, whereas film can never be re-used and even photoluminescence dosimeters need post-read treatment before they can be used again.

However, perhaps the most attractive feature of thermoluminescence for personnel dosimetry arises because LiF is a good host material. Since fluorine has an atomic number only slightly larger than oxygen, and lithium is very light, the response of LiF is very like that of human body tissue. Another good thermoluminescence material is $Li_2B_4O_7$ which has a response even closer to human tissue. Although these two materials, and many others besides, make good practical dosimeters the microscopic process which takes place in them is by no means clear. We do know that it is essential that LiF should contain magnesium ion impurities for proper thermoluminescence behaviour, and also that the efficiency of thermoluminescence is increased substantially by the addition of small amounts of second impurities such as titanium. The fact that the light is emitted in a number of different temperature peaks (figure 9.3) indicates that many defect types are involved, and it does seem likely that the different peaks correspond to defects associated with different aggregations of the Mg^{2+} impurity ions. In spite of the lack of detailed knowledge about their operating mechanism thermoluminescence dosimeters are very rapidly supplanting film and are becoming the most popular personnel dosimeter. Neutron-induced nuclear reactions involving the isotope ^6Li can be exploited to allow the use of thermoluminescence for detecting neutrons, even in the presence of other ionizing radiation.

9.2. *Information storage devices*

It is probably rare for most of us to remember that photographic film is our most widely used information storage medium, let alone that it relies for its operation on a radiation-damage process. However, it does. Light or higher energy radiation creates defects in the silver-halide grains by a photochemical process, and if enough light is used visible defects can be created directly in the film. Usually, however, the exposure is stopped when the defects are still too small to see and their size is increased later by chemical means (developing the film).

Photographic film is certainly very cheap and widely used to store visual images, but it cannot cope with the more exacting information storage requirements of electronic computers because it suffers from the two major disadvantages of being irreversible and of needing chemical

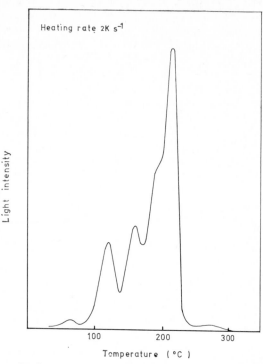

Fig. 9.3. Thermoluminescence in LiF:Mg as the crystal temperature is raised (after P. D. Townsend and J. C. Kelly, *Colour Centres and Imperfections in Insulators and Semiconductors* (Chatto & Windus, London, 1973)).

development. Yet the increasing use of electronic computers generates a great need for cheap methods of information storage, and those methods which, like film, rely on radiation effects do offer exciting possibilities. To illustrate the philosophy of these memory devices we need to look briefly at some of the basic features of information storage.

Information for computers is usually measured in bits (binary digits), with each bit having the value 0 or 1. Many of the most reliable electronic methods of storing and processing information, such as 'flip-flop' circuits, have only two possible states and make natural stores of binary information, each flip-flop storing one bit. However, a book like this one contains about $1\frac{1}{2}$ million bits of information, since it contains about 50 000 words of about 6 letters each and we could replace each letter by 5 bits by using the binary number system, putting 00001=A, 00010=B, 00011=C . . . 11010=Z. Making the information in this book available to a computer would therefore need $1\frac{1}{2}$ million flip-flop circuits! On the other hand, a single television picture also contains about 1–2 million bits of information, since the picture is composed of

174

about 600×600 picture points, each of which can have any one of about 30–50 brightness levels, so that each picture point needs 5–6 bits to define its intensity fully. In principle, therefore, a *single* photographic negative can contain as much information as the whole of this book and as much as $1\frac{1}{2}$ million electronic flip-flop circuits. This comparison illustrates the power and potential cheapness of information storage using a memory plane which is basically like photographic film, but more adaptable. It is probably worth pointing out that photographic film stores information so cheaply because it is 'geographical' in character, that is a bit corresponds to a location on a homogenous medium. This is in contrast with 'digital' memories like flip-flops, where each bit corresponds to a discrete piece of hardware.

However, computer memories must be fast as well as cheap. A modern computer can process about 100 million bits of information every second; it could read and memorize this book in one-hundredth of a second. Flip-flops using modern integrated circuit technology are a partial solution; here we are starting with a basically fast digital memory and trying to make it cheaper. An obvious alternative approach is to start with a cheap geographical memory and try to make it faster. Until now a common feature of all geographical memories has been the use of mechanical access to the required location on the continuous storage medium. Magnetic discs and tapes are the best known examples. In these memories the information is alterable, and therefore acceptable for computer use, many times cheaper per stored bit than discrete transistor circuit information storage even when the latter is cleverly integrated on a single silicon 'chip', but the dependence on the mechanical movement of the recording medium under a recording- or reading-head results in relatively long access times, of several tens of milliseconds for discs and many seconds for tape. The new feature where photographic film is used for geographical information storage is that access can be achieved very quickly, by scanning an electron beam or a light beam over the film. For photographic film itself the need for chemical development and the inability to alter the information remain major obstacles.

Many research workers have therefore sought a new kind of film for information storage, which does not need development and which can be reversed. The potential commercial rewards are very high and many possible materials have been examined. So far none has achieved widespread acceptance. However, we will look at two contenders, representing information storage technologies based on defects where the access is by a light beam and an electron beam respectively.

Photochromic calcium fluoride

In Chapter 5 we discussed the optical properties of F centres in alkali halides. These defects are halogen ion vacancies with a single electron trapped in each, and they give rise to strong optical absorption bands. Related defects called F_A centres also form in alkali halides. These are also halogen ion vacancies with single trapped electrons, but instead of having six equivalent alkali ions in the nearest-neighbour positions they have only five normal lattice cations and one foreign alkali cation. For example, an F_A centre in KCl could have five K^+ ions and one Na^+ ion around the Cl^- ion vacancy.

F_A-like centres can also be created in CaF_2, with one of a number of rare-earth ions playing the role of the foreign alkali ions. It therefore seems sensible to call them F_{RE} centres. Like F_A centres in alkali halides they also have strong absorption bands in the visible and near ultra-violet parts of the spectrum, but when photons are absorbed in these bands the centres are usually ionized, that is, they lose electrons into the conduction band of the crystal and become F_{RE}^+ centres.

The F_{RE} centres remain ionized (as F_{RE}^+) only if other defects can be provided to trap the electrons released. In CaF_2 these other centres can conveniently be trivalent rare-earth ions. Thus in CaF_2 containing F_{La} centres and La^{3+} ions the ionization process would be represented by the chemical equation

$$F_{La} + La^{3+} \xrightarrow{\quad h\nu_1 \quad} F_{La}^+ + La^{2+} \tag{9.1}$$

The new state of the crystal is not fully stable, but it does last for about a day. However, much faster return to the original state can be achieved using the strong absorption band due to La^{2+} centres in the visible part of the spectrum. This band is at a different wavelength from the F_{La} band and again the light causes the absorbing centre (La^{2+}) to be ionized, so that eqn. (9.1) is reversed:

$$La^{2+} + F_{La} \xrightarrow{\quad h\nu_2 \quad} La^{3+} + F_{La} \tag{9.2}$$

These two processes allow information to be written on a crystal of CaF_2 containing La^{3+} and F_{La} centres. The F_{La} band can be used for writing and the La^{2+} band for erasing the information. All the four defects involved, F_{La}, F_{La}^+, La^{2+}, La^{3+}, are immobile in the material, and only electrons are moved around in the writing–erasing cycle. Cycling can therefore be carried out as often as necessary, without any degradation.

Reading can be carried out with light in either the F_{La} or La^{2+} bands, but in those cases the reading would destroy the information. However, the F_{La} centre actually has two fairly strong optical absorption bands. The first is at about 400 nm and ionizes the centre, as we have described, but there is a second band at about 720 nm. Although the absorption at 720 nm is also fairly strong this light does not ionize the centre or change its state in any way. It can be used to decide the state of the

176

crystal without changing it at all. All this behaviour of CaF$_2$ containing F$_{La}$ and La^{3+} centres is summarized in figure 9.4. This material, then, behaves rather like photographic film which does not need chemical processing and is reversible. We can write information into the crystal using light at 400 nm, and in doing so we significantly change the optical properties of the crystal at 720 nm. The content of the memory plane can be read without causing any change by using light at 720 nm, and can be altered when necessary with light at 550 nm. The writing and erasing processes are very nearly as efficient as is theoretically possible and the material does not degrade when it is cycled from one state to the other and back many times.

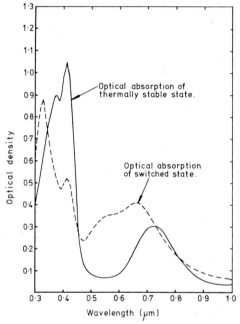

Fig. 9.4. The optical absorption spectra of F$_{La}$ centres and La^{2+} centres in CaF$_2$ (after R. C. Duncan, *RCA Review*, **33**, 248 (1972)).

It is nearly the ideal optical information storage medium but not quite. One disadvantage which we have already mentioned is the instability of the switched state, which lasts only for a day for the lanthanum centres and up to a week for cerium. Secondly, it is difficult to achieve high concentrations of F$_{RE}$ and RE^{2+} centres, which means that relatively thick specimens must be used to give reasonable optical density changes. A CaF$_2$ memory plane would have to be accessed with a focused light beam, which means that a thick plane increases the area that must be used to store a single bit (figure 9.5). The number of bits

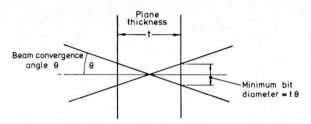

Fig. 9.5. The relationship between the thickness of an optical memory plane and the minimum diameter used to store a single bit.

which can be stored in a given area is therefore reduced and the cost per bit increased; it seems likely that an area of about 200 mm × 200 mm would have to be used for 1 million bits, and this is rather large.

A third obstacle to the use of this photochromic material as an optical memory plane comes from the seemingly attractive feature of three distinct optical absorption bands for write, erase and read processes. For all optical memories lasers are the only light sources which are sufficiently intense to carry out the read or write operations quickly enough for computer use. Although laser beams can be deflected to access an array of bits, by electro-optic or acousto-optic devices, it is as yet impossibly difficult to arrange to have two or three different lasers scanned over the same array of stored bits. Nevertheless, memory applications of F_{RE} centres in CaF_2 may well be found.

Inhibitable cathodoluminescent phosphors

One method of storing information in crystals by means of radiation-damage processes has been examined at Harwell. It is very different in concept from the optical memory based on CaF_2 photochromics in that an electron beam rather than a light beam is used to access the memory. The basic phenomenology is fairly simple, and can best be explained by using as an example inhibitable cathodoluminescence in KI:Tl (potassium iodide containing thallium impurity ions).

We saw in Chapters 6 and 7 that high energy electrons first create electron–hole pairs in insulators, and that in KI:Tl electrons can be quickly trapped by Tl^+ ions to form $Tl°$ atoms. It turns out that at room temperature about half the holes created with the electrons find these $Tl°$ atoms very quickly (within 200 ns) and reconvert them to Tl^+. In this recombination process light is emitted in a broad band centred at 420 nm, the character of the light being partly determined by the Tl atoms and partly by the host lattice. Incidentally, the other half of the electron–hole pairs also recombine at Tl^+ ions, and also cause emission of 420 nm light, but this occurs rather more slowly (in about 200 μs at room temperature).

We also saw in Chapter 7 that electron–hole recombination can cause photochemical radiation damage in alkali halides, creating F centres and interstitial atoms. The latter normally aggregate into large clusters at room temperature, but when the crystal contains large numbers of impurity atoms the interstitial atoms tend to combine with these rather than with one another. It turns out that the complexes of Tl^+ ions and interstitial atoms behave very differently from the bare Tl^+ ions. Electron–hole recombination does still take place at the complexes more or less as it does at the bare Tl^+ ions, but no light is emitted in the process.

As a result KI:Tl, which initially luminesces very brightly at around 420 nm under electron irradiation, slowly loses its efficiency if continuously excited, because the electrons create interstitial atoms which are trapped at the Tl^+ activator ions and prevent their luminescing. This fall in light output is very large, being exponential with electron irradiation dose as shown in figure 9.6 and covering more than two

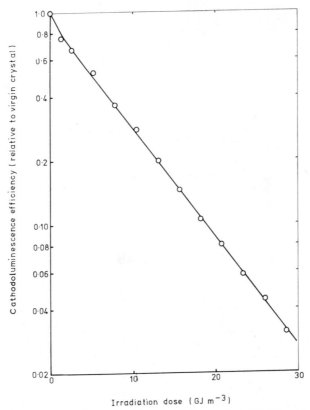

Fig. 9.6. The fall in cathodoluminescence efficiency of KI:Tl with electron irradiation dose.

179

orders of magnitude. A part of a crystal which has been inhibited in this way can easily be distinguished from those parts which have not by virtue of this large difference in luminescence efficiency, figure 9.7 (a). In a memory device a fairly intense electron beam can be used to *write* a pattern of information which can then be *read* with the same electron beam at much lower current. Defects are created in the reading process just as during writing, so that reading a location a large number of times will eventually convert it into the inhibited or dim state. However, because the electron dose required for reading is so small (as little as 5 pJ is adequate) each location can in fact be read a large number of times (10^5–10^8) before this happens.

(a) Electron beam inhibition

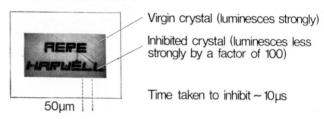

Virgin crystal (luminesces strongly)

Inhibited crystal (luminesces less strongly by a factor of 100)

50μm

Time taken to inhibit ~ 10μs

(b) Electron beam restoration

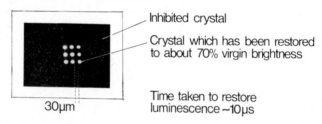

Inhibited crystal

Crystal which has been restored to about 70% virgin brightness

30μm

Time taken to restore luminescence ~10μs

(c) Storage density

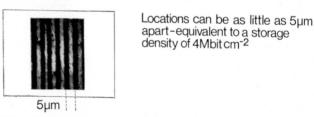

Locations can be as little as 5μm apart – equivalent to a storage density of 4 Mbit cm^{-2}

5μm

Fig. 9.7. The phenomenology of the Harwell memory plane.

So far we have not mentioned an erase process. The defects created by electron irradiation can, of course, be annealed out by simple heating and then the luminescence is restored. However, what makes the KI : Tl type of memory plane interesting is that if the electron beam used for writing and reading is made very intense the Tl^+-interstitial atom complexes are destroyed by the electron beam, and the crystal appears to return to the virgin state, once again having a high luminescence efficiency. Although this return to the virgin state is certainly due to the defect annealing caused by the heating effect of the intense electron beam, the speed at which the process occurs is increased by many orders of magnitude over purely thermal annealing. The likely reason is that the ionization caused by the electron beam excites defects into states where they break up and move around much more readily than in the unexcited state, so that electron beam annealing in KI:Tl is a million times faster than thermal annealing at the same temperature. This return of luminescence in a previously totally inhibited crystal is illustrated in figure 9.7(b) and gives the memory its *erase* process.

A memory system using this plane would probably be like that shown in figure 9.8, and it could contain 10^6–10^7 bits per tube with fairly

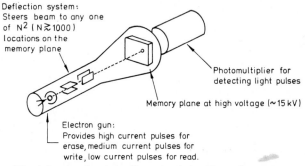

Deflection system:
Steers beam to any one
of N^2 ($N \gtrsim 1000$)
locations on the
memory plane

Photomultiplier for
detecting light pulses

Memory plane at high voltage (~ 15 kV)

Electron gun:
Provides high current pulses for
erase, medium current pulses for
write, low current pulses for read.

Fig. 9.8. A schematic diagram of the Harwell memory.

simple electron beam deflectors and perhaps 10^8–10^9 bits if a more complex electron beam deflection system were used. One important factor in the cheapness of the Harwell memory is the fact that the memory plane itself is homogenous and easy to fabricate. Another factor is that the electron beam determines its own set of memory locations, which are automatically the same for read, write and erase, because an electron beam of the same energy is used for each. It does not much matter if the array of locations is slightly distorted from a perfect square lattice, nor does the array have to be positioned very precisely with respect to the memory plane itself, and this tolerance makes construction easier and cheaper. There are other important advantages such as the very large number of bits (10^6–10^7 for simple electron optics and 10^8–10^9 for more complex tubes) which can be reached through a single input and output channel.

One obstacle in the path of the widespread application of this memory

181

is the degradation of a particular storage location which occurs when it is cycled through the dim and bright states many times. The materials available at present have a limited life of only 10^4 reversals, but better ones may become available in the future, and in any case the degradation can be removed thermally, which could be adequate for some purposes.

Dark trace displays

As we saw in Chapter 5 the most spectacular properties of defects in insulating crystals are their optical ones; in fact, defects in insulators are often called colour centres. The best known and most widely studied colour centres are F centres in alkali halides, and it is fitting that they were the first to find an application, during the second world war, in dark trace display tubes. The problem that the scientists faced at that time was to find a way to present radar to the operators. The most common form of radar has a large aerial which is rotated slowly around a vertical axis. When the aerial is pointing at a target ship or aeroplane the electronics can detect it and measure its range, but once the aerial has rotated past the target the range and bearing information cannot be generated again for several seconds, until the aerial has made a complete revolution. Usually a simple cathode-ray tube is arranged to scan in synchronism with the aerial, as in figure 9.9. The disadvantage of this is that the decay of the bright spot corresponding to a target is fairly rapid even with a long persistence phosphor, and this makes the display unpleasant to view even in a dark room, and often impossible to view in daylight.

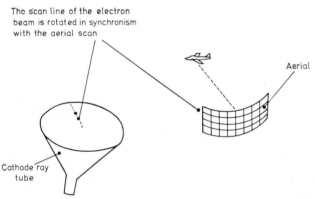

Fig. 9.9. The basis of the position indicator.

In the dark trace display of World War II a cathode-ray tube was still used, but the phosphor was replaced by a thin layer of KCl. Instead of generating light the electron beam created F centres in the KCl layer, which were visible from the front as a dark trace, by virtue of their intense optical absorption bands in the visible region. The pattern of F centres did decay slowly because of the bleaching effect of the viewing light, but the decay can take place over many seconds, and therefore provides the display with the built-in information storage needed for radar. The dark-trace display tube had a secondary and important advantage over a simple cathode-ray tube, in that it could be seen as easily in bright sunlight as in a dimly lit room, which is particularly attractive for displays in ships or aircraft.

In the end the KCl 'Skiatron', as it was called, was not very successful. This was partly because insufficient optical contrast was available and partly because material degradation occurred. However, the need in radar remains and new applications have appeared, such as for displays driven by computers where rapid refreshment of the display can be very expensive and sometimes impossible. Because of this, new materials have been developed, notably complex aluminosilicates containing halide ions called sodalites (for example $Na_6Al_6Si_6O_{24}:2NaBr$), which provide much better performance. None of these new materials has yet achieved the right mixture of properties to ensure a really wide use for dark-trace display tubes, but small numbers are manufactured. A typical dark-trace display is shown in figure 9.10.

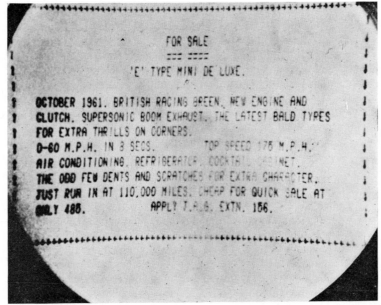

Fig. 9.10. A dark-trace display (after P. A. Forrester, D. J. Marshall, S. D. McLaughlan and M. J. Taylor, *The Radio and Electronic Engineer,* **40,** 17 (1970)).

9.3. *Other useful radiation damage effects*

In solid-state radiation dosimetry radiation damage is used simply as a measure of the total amount of radiation received. Without doubt this is a very useful radiation-damage effect, but the irradiated dosimeter material itself is not of any interest except for the information it contains. This is also true of the irradiated material in information storage devices. There are some cases, however, where irradiated material is more valuable than unirradiated, and where irradiation is carried out to make this desirable transformation. One important use of radiation in this way which we will not discuss is in sterilization of medical supplies and also of food. Radiation does kill micro-organisms and viruses very effectively but these are hardly simple solids. For a proper discussion of these processes the reader is referred to *Biological Effects of Radiation.*

Polymer impregnation of wood

Even today wood is widely used for structural and decorative purposes. It can be worked easily with simple steel tools and is fairly strong and tough. Although it does need protection from the weather it is basically a fairly resistant material and is still a favourite of the building industry. Its attractive appearance and relatively low price further ensure its widespread use. However, its properties are very variable from one specimen to another of nominally identical wood, depending for example on whether the specimen is heartwood or sapwood and on the conditions of growth and seasoning. A further major problem with wood arises from its porosity, which can represent more than three-quarters of the total volume and allows the wood to take up moisture when the atmosphere is wet and release it when dry. The swelling and contraction which follow the absorption and release of water can cause volume changes of up to 50 per cent, and often result in warping or splitting.

One technique for improving wood has been developed at a number of radiation laboratories, including Harwell, and involves impregnating the wood with a suitable monomer and then γ-irradiating it to convert the monomer to polymer form. A favoured monomer is a mixture of styrene and acrylonitrile which have the following chemical structures:

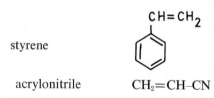

styrene

acrylonitrile $CH_2{=}CH{-}CN$

When a 60:40 mixture of these two compounds is irradiated with a dose of about 2 megarad a hard tough polymer results. In treating wood

184

the specimens are first evacuated, to remove all the air from the pores, and then impregnated with the monomer mixture under pressure. The monomer mixture is then polymerized by irradiation in a ⁶⁰Co source.

Wood polymer composites have many improved properties, compared with untreated wood. They are harder, stronger and much more resistant to surface damage, but they can still be sawn, planed and drilled. The plastic content of the composite allows surface polish to be obtained without the use of any waxes, and gives the surface a good resistance to water and many other chemicals. The reduction in water uptake can be spectacular, as shown in figure 9.11, which greatly improves the dimensional stability. The major use at present is for special flooring and walkways, where the high abrasion resistance and insensitivity to chemical attack are of primary importance, but the number of applications is growing steadily.

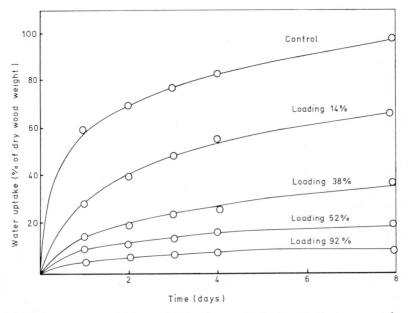

Fig. 9.11. Water uptake into sycamore impregnated with 60:40 styrene:acrylonitrile (after P. R. Hills and D. J. McGahan, *Atom*, No. 176, 146 (1971)).

The mechanism of the radiation polymerization of molecules such as styrene or acrylonitrile is actually very complicated, but the essential features are as follows. By a photochemical process the radiation breaks up one of the monomer molecules. Let us suppose that it removes a hydrogen atom. The hydrogen atom then starts a chain polymerization process in which at each stage a free radical is formed which can initiate further polymerization:

$$H \cdot + CH_2 = CH\text{-}R \longrightarrow CH_3\text{-}\overset{\displaystyle R}{\overset{\displaystyle |}{C}}H \cdot$$

$$CH_3\text{-}\overset{\displaystyle R}{\overset{\displaystyle |}{C}}H \cdot + CH_2 = \overset{\displaystyle R}{\overset{\displaystyle |}{C}}H \longrightarrow CH_3\text{-}\overset{\displaystyle R}{\overset{\displaystyle |}{C}}H\text{-}CH_2\text{-}\overset{\displaystyle R}{\overset{\displaystyle |}{C}}H \cdot$$

and so on. In these chemical equations the symbol R can be $-CN$ or C_6H_5 for the acrylonitrile/styrene mixture.

Polymerization can, of course, be initiated chemically as well as by irradiation but this always involves the heating of the system, and radiation polymerization is much more convenient for wood composite formation.

Ion implantation

We saw in Chapter 4 how impurities normally dominate the electrical properties of semiconductors, and how this allows the production of semiconductor diodes and transistors. These solid-state devices have replaced thermionic valves in most applications which do not need extremely high voltages or powers, and this has been a revolution in electronics. An even more dramatic outcome of the replacement of valves by transistors has been the development of integrated circuits during the past 5–10 years. In these, large numbers of transistors, with resistors, connections and sometimes capacitors, are incorporated onto a single piece of silicon (usually called a chip), so that wiring costs and material costs are minimized, while stray capacitances are also very small. Many thousands of transistors on a single chip of silicon are now commonplace (1974), while the most advanced circuits available contain several tens of thousands. The advantages of large-scale integration are primarily low cost, low power consumption and high performance, and all these are achieved because large numbers of components are packed closely together on a single chip of silicon. If this development is to continue still further, with more and more components per unit area of silicon, methods must be found for placing the crucial impurities in the silicon more and more precisely.

In most present-generation silicon devices the donor or acceptor impurity atoms are diffused into the silicon at high temperature; typically a diffusion time of a few tens of minutes at 1000–1300°C is necessary. There are therefore problems arising directly from the lack of sharpness of the edge of the diffused layer, which makes very sharp p–n junctions difficult to obtain. However, possibly even more important than this is the fact that a number of diffusion, passivation (growing onto the silicon a layer of protective SiO₂ film) and metallization steps are necessary to make a complex integrated circuit. Each diffusion step requires heating to a high temperature and it is nearly impossible to prevent unwanted interactions between them; atoms put in at one time may be moved by a subsequent diffusion step. A way round this problem is to use an ion accelerator to implant the required impurity atoms directly where they are needed in the silicon. This can be done at room temperature if required, although then the radiation damage caused by the implanted ion tends to dominate the effect of the ion itself. At 600–700°C, however, the radiation damage anneals out, but the temperature is still far too low for significant diffusion of the impurity ions. The implantation energy need not be large; figure 9.12 shows how even fairly low energy implantation can result in useful penetration, so that ion implantation equipment can be relatively cheap.

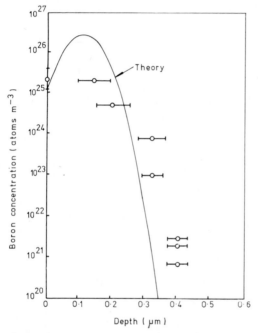

Fig. 9.12. The distribution of electrically active boron ions implanted at 30 keV into silicon (after J. W. Mayer, L. Eriksson and J. A. Davies, *Ion Implantation in Semiconductors*, Academic Press, New York (1970)).

One example of where ion implantation is used because of the sharpness which can be achieved in junctions is in the manufacture of 'hyperabrupt' voltage-controlled capacitor diodes. These are called Schottky diodes and have the structure shown in figure 9.13. Nowadays they are formed by ion-implanting a layer of impurity ions into the

187

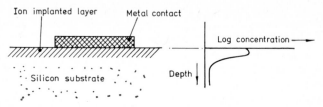

Fig. 9.13. Hyperabrupt Schottky barrier diodes formed by ion implantation.

silicon and evaporating on the metal. The capacitance of the junction varies with voltage, it would vary with $V^{-\frac{1}{2}}$ if the shaded layer in figure 9.13 were very thick and fairly uniformly doped, but it varies much faster for a sharp profile such as can be produced by ion implantation. These variable capacitance (varactor) diodes have applications in automatic frequency control circuits and in frequency multiplication, where they can give multiplication at high efficiency without need for a d.c. power supply. The more rapidly the capacitance varies with voltage the easier multiplication is to achieve; hence the use of implanted varactor diodes.

Many other examples of the advantages of ion implantation can be cited but perhaps the most important application is the gate-masked ion-implanted MOSFET. Figure 9.14 (a) shows the structure of a conventional diffused MOSFET. The substrate is relatively pure silicon and

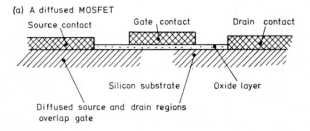

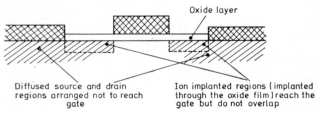

Fig. 9.14. Comparison of diffused and ion-implanted MOSFETs.

the source and drain regions can be either p or n type. When a voltage is applied between the source and drain contacts a current flows along the oxide–semiconductor interface which can readily be controlled by varying the voltage on the gate electrode. Because the oxide layer separates the gate from the silicon the resistance between the gate and the source or drain is usually very high (10^{10}–10^{12} Ω). However, the gate to source or drain capacitances can be fairly large, primarily because the gate has always to be made several μm wider than is really

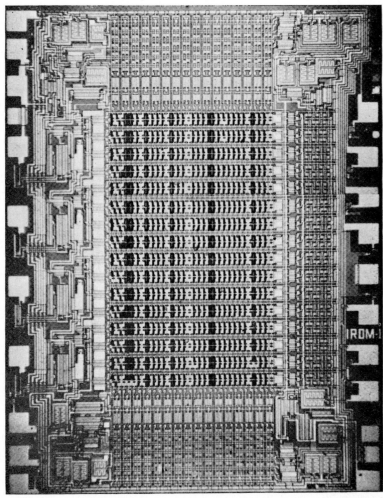

Fig. 9.15. A photomicrograph of an ion-implanted 2048-bit read-only memory (after J. W. Mayer, L. Eriksson and J. A. Davies, *Ion Implantation in Semiconductors*, Academic Press, New York (1970)).

189

necessary to span the gap between source and drain, just to allow for errors in manufacture and to ensure that the gap is always completely spanned. An ion-implanted MOSFET (figure 9.14 (*b*)) is made initially in the same way as a diffused one, except that the gate is made deliberately too narrow. Ions are then implanted *through* the thin oxide layer as shown. They do not reach the silicon under the gate because they are stopped in the metal, and they do not significantly change the electrical properties of the metal. They do, however, extend the source and drain regions exactly to the gate, thus giving the minimum possible capacitance consistent with reliable operation. A reduction of a factor of 5 in capacitance can be achieved in this way, and as a result more complex circuits can be made on a given chip with a given performance. Figure 9.15 shows a photograph of a 2048-bit ion-implanted MOS read-only memory, on a chip 3 mm × 4 mm. Many modern micro processors, 'computers on a chip', are made at least in part by ion implantation.

General summary

We have tried to show in this chapter how defects produced by irradiation can be harnessed and used to man's advantage. In the book as a whole we have first tried to emphasize the importance of defects in solids, turning as they do the perfect crystals which hardly ever exist even in the research laboratory into the real solids which make up the world around us. Their importance here can scarcely be overestimated, they give steel its strength, the colour to stained-glass windows and paintings, and they allow modern radios and computers to work. Then we have looked at how solids interact with radiation, at how this determines the character of the world we see and how it influences the design of the nuclear and thermonuclear reactors on which we may depend so heavily for warmth and light in the future. Solids, defects and radiation effects provide just another example of how science is composed of closely interlocking areas of knowledge. They demonstrate again that there is an unbroken pathway from the microscopic processes of nature, such as collisions between atoms, to nuclear reactors and computers on a chip which, for good or bad, are part of our advanced civilization.

SELECTED BIBLIOGRAPHY

General Solid State Physics and Chemistry:
D. Tabor, *Gases, Liquids and Solids,* Penguin Books, London, 1969.
C. A. Coulson, *Valence,* Oxford University Press, 1952.
B. R. Jennings and V. J. Morris, *Atoms in Contact,* Clarendon Press, Oxford, 1974.
C. Kittel, *Elementary Solid State Physics,* John Wiley and Sons, New York, 1962, and *Introduction to Solid State Physics,* 3rd edition, John Wiley and Sons, New York, 1967.
F. C. Brown, *The Physics of Solids,* W. A. Benjamin, New York, 1967.
A. J. Dekker, *Solid State Physics,* Macmillan, London, 1962.
R. J. Elliott and A. F. Gibson, *Introduction to Solid State Physics and its Applications,* Macmillan, London, 1974.
A. G. Guy, *Introduction to Materials Science,* McGraw-Hill, New York, 1972.
L. Solymar and D. Walsh, *Lectures on the Electrical Properties of Materials,* Clarendon Press, Oxford, 1970.
N. B. Hannay, *Solid State Chemistry,* Prentice-Hall, New York, 1967.

Defects and their influence in the properties of solids:
H. G. Van Buren, *Imperfections in Crystals,* North Holland, Amsterdam, 1960.
A. H. Cottrell, *Dislocations and Plastic Flow in Crystals,* Clarendon Press, Oxford, 1953.
B. Henderson, *Defects in Crystalline Solids,* Edward Arnold, London, 1972.
P. D. Townsend and J. C. Kelly, *Colour Centres and Imperfections in Insulators and Semiconductors,* Chatto and Windus, London, 1973.
F. A. Kroger, *The Chemistry of Imperfect Crystals,* North Holland, Amsterdam, 1964, revised into three volumes, 1973.
N. N. Greenwood, *Ionic crystals, Lattice Defects and Non-stoichiometry,* Butterworths, London, 1968.
L. A. Girifalco, *Atomic Migration in Crystals,* Baisdell, New York, 1964.
P. Kofstad, *High Temperature Oxidation of Metals,* John Wiley and Sons, New York, 1966.
A. Kelly, *Strong Solids,* 2nd edition, Oxford University Press, 1972.

Radiation effects
D. S. Billington and J. H. Crawford, *Radiation Damage in Solids,* Princeton University Press, 1961.
R. Strumane, J. Nihoul, R. Gevers and S. Amelinckx (eds), *The Interaction of Radiation with Solids,* North Holland, Amsterdam, 1964.
J. W. Corbett, *Electron Radiation Damage in Semiconductors and Metals,* Solid State Physics Supplement 7, Academic Press, New York, 1966.
L. T. Chadderton, *Radiation Damage in Crystals,* Methuen, London, 1965.
M. W. Thompson, *Defects and Radiation Damage in Metals,* Cambridge University Press, 1968.
J. W. Mayer, L. Eriksson and J. A. Davies, *Ion Implantation in Semiconductors,* Academic Press, New York, 1970.
G. Dearnaley, J. H. Freeman, R. S. Nelson and J. Stephen, *Ion Implantation,* North Holland, Amsterdam, 1973.
R. S. Nelson, *The Observation of Atomic Collisions in Crystalline Solids,* North Holland, Amsterdam, 1968.
S. Amelinckx, B. Batz and R. Strumane (eds), *Solid State Dosimetry,* Gordon and Breach, New York, 1969.

INDEX

195

Radiothermoluminescence 172ff
Random walk 39ff
Range of particles 112, 119
Reactor
  fast 150
  fission 149ff
  fusion 154, 161
  irradiation 132, 149ff
  thermal 150
Rectifier 49
Reflection
  by dielectrics 78
  by metals 83ff
Refractive index 75
Relativistic electron collisions 126
Repulsive energy 7ff
Resonant scattering (of phonons) 60
Reverse bias 49
Rutherford collisions 131ff

Scattering of charged particles 112ff
Schottky
  defects 27, 36ff, 42ff
  diodes 187
  pair 38, 52, 54
Screened coulomb potential 117ff
Second law of thermodynamics 36
Self-trapped hole 139ff
Self-trapping 50, 139ff
Semiconductors 19ff, 47ff
Silver-activated glass 170ff
Skiatron 183
Skin depth of metals 84
Slip 69ff
Slip plane 28, 71
Sodalite 183
Solar cells 164
Stacking fault 31ff
Stationary states 3
Stoichiometry 34
Stored energy 154ff
Straggling of charged particles 112ff

Stripping of projectiles 115ff
Styrene 184
Superconducting materials 161ff
Surface energy 67ff
Swelling 155ff

Tensile stress 66ff
Thallium centres 139ff, 178ff
Thermal annealing 145
Thermal conductivity 26, 55ff
Thermal expansion 26, 43
Thermal reactor 150
Thermoluminescence 172ff
Thomson scattering 101ff

U centres 81
Ultraviolet irradiation 142
Unit cell 10, 21

Vacancy 27, 29, 144ff
Vacuum level 137
Valence band 21ff
Valency 5
Varactor diodes 188
Voids 32, 157ff

Wave-function 6
Wavelength
  electron 94
  neutron 92
  X-ray 43, 91
Wiedemann–Franz law 56
Wigner release peak 155
Work hardening 72

X-ray diffraction 43, 91ff
X-ray fluorescence 138
X-ray generators 109

Yield stress 66

Zero-point energy 24